MAVA Math: Enhanced Skills

Marla Weiss

MAVA Books and Education Company
www.mavabooks.com

authorHOUSE®

AuthorHouse™
1663 Liberty Drive
Bloomington, IN 47403
www.authorhouse.com
Phone: 1 (800) 839-8640

Published by AuthorHouse 12/10/2015

ISBN: 978-1-5049-6778-5 (sc)

Print information available on the last page.

Any people depicted in stock imagery provided by Thinkstock are models,
and such images are being used for illustrative purposes only.
Certain stock imagery © Thinkstock.

This book is printed on acid-free paper.

Because of the dynamic nature of the Internet, any web addresses or links contained in this book may have changed
since publication and may no longer be valid. The views expressed in this work are solely those of the author and do
not necessarily reflect the views of the publisher, and the publisher hereby disclaims any responsibility for them.

CONTENTS

Page	Topic

NOTES

Books By Marla Weiss

Available at AuthorHouse and most online bookstores

FICTION

School Scandalle

School Scoundrelle

MATH WORKBOOKS

MAVA Math: Number Sense

MAVA Math: Number Sense Solutions

MAVA Math: Grade Reviews

MAVA Math: Grade Reviews Solutions

MAVA Math: Middle Reviews

MAVA Math: Middle Reviews Solutions

MAVA Math: Enhanced Skills

MAVA Math: Enhanced Skills Solutions

NONFICTION (available 2016)

How To Finally Fix Math Education

Available at www.terrapinlogo.com

COMPUTER PROGRAMMING WORKBOOKS

Go, Logo!

Go, Logo! Solutions

An Important Message To Teachers, Parents, and Students

What are Enhanced Skills?
All *MAVA Math* workbooks have two-word titles: *Number Sense*, *Grade Reviews*, and *Middle Reviews*. *Enhanced Skills* keeps that format while indicating that the combined lessons are of a more advanced nature, suitable for students who wish to study math in depth or to enter math competitions.

What is this book's curriculum?
The detailed specificity of the Contents acts as a curriculum guide. This book includes a blend of basic skills as well as enriched material.

What is the grade level of this book?
While this book aims at grades six through eight, some students may begin earlier or later.

How many years should students take to complete this book?
Mathematically gifted students may complete the book in grades 6 and 7, finishing any remaining pages concurrently with Honors Algebra I in 8th grade. Other students could complete the book in 3 years.

Why do topics appear in alphabetical order?
Presenting the topics in alphabetical order offers maximum flexibility for teachers and students who may construct their own desired progression of lessons. Alphabetical order also permits students to easily supplement classroom work or contest preparation.

Does this book use a developmental approach?
Yes. Within each of the 79 topics, lesson 1 precedes lesson 2 developmentally. With AGE PROBLEMS for example, students should learn how to solve problems linearly prior to constructing charts. However, for the topics with many lessons, slight reordering may make sense for some teachers and students.

Are these topics and problems all that exist in advanced middle school math?
No. While this book is very comprehensive, another 426 pages still would not exhaust the rich variety of possible exercises.

Does this book's curriculum vary dramatically in any way from those typically seen?
Yes. This book's breadth, depth, and quantity of problems are not commonly found in one comprehensive volume.

Was this book field-tested?
Yes. Many problems in this book were used as part of comprehensive worksheets written by Marla Weiss for classroom settings. This material, unified for the first time in this book, yielded students who loved math, performed high on standardized tests, and earned countless awards at math competitions.

Does MAVA Math offer another middle school level workbook?
Yes. *MAVA Math: Middle Reviews*, intended to supplement the daily textbook, is designed for all middle school students. The accompanying answer book is *MAVA Math: Middle Reviews Solutions*.

May this book be used as the sole textbook?
Yes. This book in draft form was used as the sole textbook, supplemented with contest materials, for grades 6 and 7. Those students excelled in math.

Should certain pages of certain topics be done before others?
Yes. For example, students should know DIVISIBILITY 1 (rules) before FRACTIONS 1 (simplifying) or FRACTIONS 2 (multiplying). This book trusts teachers to order math.

Does this book prepare for future math instruction?
Yes. This book provides a solid foundation for high school math, including algebra skills, algebra word problems, plane geometry, and space geometry.

Should students use a calculator with this book?
Students should use a calculator only when absolutely essential. This book encourages practice of number sense throughout, not only on the pages titled MENTAL MATH.

What are the different types of MENTAL MATH?
Mental math may mean: 1. responding to an oral question; 2. doing all work in one's head for a written problem; and 3. doing minimal written work for a written problem.

Does this book cover a complete course in Algebra I?
No. This book covers most topics in pre-algebra and some topics in algebra but is not in any way intended to cover a full Algebra I course.

Why do some answers abbreviate words?
Math class is not an opportunity to teach language arts, whether spelling, handwriting, or composition. Attaching verbal skills to the study of math slows students both in daily work and annual progress. Ideally, students will learn to express themselves mathematically. However, mathematicians write in a specific style, totally different from English essays. Most abbreviations used in this book are found on pages 427–428.

Why do some geometry problems use upper case as well as lower case abbreviations?
When a problem involves two circles, use capital R for the radius of the larger one and lowercase r for the radius of the smaller one. Moreover, use B for the larger base of a trapezoid and b for the smaller one. Clarity of notation leads to improved focus and accurate answers.

Why are the answers to some measurement problems a number without a label following?
Consider the question: How far do you live from school? The answer could be 2 miles or 2 turtle steps. A label is needed for accuracy. Now consider the question: How many miles do you live from school? The answer may be 2 without ambiguity because a label, namely miles, is built into the question. Requiring a label at all times is unnecessary.

Should all improper fractions be converted to mixed numbers?
No. The term "improper fraction" is a misnomer. Converting from a fraction greater than one to a mixed number is a valuable skill, but it need not always be done. For example, as a solution to an equation, the fraction is better because it may be substituted directly to check its validity. However, measurements are best as mixed numbers. For example, one and three fourths cups flour is usually more helpful than seven fourths cups.

How does a student best learn problem solving?
While the ultimate goal is to solve a variety of problems in random order, students need to learn one problem type at a time. After mastery, then problems may be jumbled.

Can cumulative review be built into this book?
Yes. Students do not have to complete a page once starting it. Returning to a page at a later time builds in cumulative review.

Why do some problems have charts and diagrams pre-drawn while others do not?
At the advanced level of this book, students need to be able to draw their own diagrams and construct their own charts. However, in some problems, a diagram shows information that is not in the text.

Which are more valuable–fractions or decimals?
Decimal math may be easily done on a calculator. Fractions are more important in higher math. For example, a student who does not understand how to add 1/2 + 1/3 cannot possibly add 1/x + 1/y (diagonal fraction lines for ease of typing only). Furthermore, fractions yield an exact answer when decimals sometimes yield an approximation.

Do some problems or skills have more than one method of solution?
Yes. To find the perimeter of a rectangle, should one add the length and width and then double, or should one double each measurement and then add? To find the slope of a line given 2 points, which point should be considered the 1st point and which the 2nd? Students should understand both methods and decide based on the numbers in the problems, always seeking fast and accurate calculation. The inherent richness and beauty of math yield multiple approaches to many problems. PERCENTS 5 (x is y% of z) is just one of many pages that show multiple methods. The WLOG method is another example.

Why do some of the 79 topics have "see" following them?
Due to the richness of math, placing a concept or skill into a category may be difficult. For example, integers may be even or odd, and equations may contain exponents. Similarly, a problem about the percent change in the area of a rectangle covers three topics. The placement is often arbitrary. Thus, the "see" references guide the user.

Why does this book use two different fonts?
All problems appear in the Arial font. All work, answers, and comments appear in the Chalkboard font. Final answers are in Chalkboard Bold.

Why are negative, opposite, and subtraction signs all represented by the same symbol?
Some math texts use both a smaller, higher line and a longer, mid-level line. Because all three signs operate equivalently, this book uses just the longer, mid-level line for simplicity.

Why are the decimal points bold?
Some students do not see decimal points in normal font. Similarly, some students do not write decimal points darkly enough. A happy medium exists between a light dot and a wart.

How can one learn more about various problem types and solution methods?
The website www.mavabooks.com offers short videos giving instruction on various topics. More will be added as time permits.

MAVA Math: Enhanced Skills Copyright © 2015 Marla Weiss

Why are equilateral triangles also labeled isosceles?
The definition of isosceles triangle is a three-sided polygon with at least 2 congruent sides. Therefore, an equilateral triangle is isosceles, but the converse is not true.

Why is functional notation used in situations that do not involve functions?
Functional notation is precise and concise. For example, P(even) is neater than writing "probability of tossing an even." Similarly, GCF(35, 49) is neater than writing the "greatest common factor of 35 and 49."

What does the word "unit" mean?
Unit is a general term. Regarding distance, "unit" may mean many different measurements such as inches, feet, miles, or centimeters. Understanding that the label must be square units for area and cubic units for volume is more important than what the actual unit is. By using the generic "unit," students may focus on specific skills.

Why does this book list over 500 vocabulary words?
Students cannot do math problems without understanding the words contained therein. Unfortunately, many math words have multiple meanings. Consider base–e.g., base of a triangle, base two arithmetic, and a number (base) raised to a power (exponent). Students learn math vocabulary when they continually hear the words used correctly.

May parents help with the pages?
Students who receive continual math help from their parents often show less growth than students who learn to work independently. Moreover, most parents have forgotten math or don't know the best ways to approach many problems. Parents should only monitor a child's work, determining weak areas needing further help.

May students and teachers write diagonal fraction lines?
Never! For correct fraction work, students must clearly see numerators and denominators, only accomplished by writing horizontal fraction lines. This book occasionally uses diagonal fraction lines for ease of typing only.

Should students memorize all of the Pythagorean triples and prime numbers at the end of this book?
No. The lists are for reference. However, math memorization should not stop with the four whole number operation facts. Further memorization, which will speed work, may include primes, Pythagorean triples, formulae, perfect squares and cubes, and square roots of non-squares (root 48 is 4 root 3).

What happens if MAVA Math: Enhanced Skills or MAVA Math: Enhanced Skills Solutions contains an error?
Both books were thoroughly proofed. However, any needed corrections will be posted on www.mavabooks.com. Please send a concise and precise e-mail to info@mavabooks.com if a correction does not address your concern.

Should math be fun?
Of course, math should be fun. However, teaching math solely as a game does not lead to growth. Students who study rigorous math truly learn math. Understanding in turn leads to natural enjoyment. Competence is pleasurable.

Absolute Value 1

Evaluate, working down. Show one line before the answer.

1. $\|3 - 7\| - 3 + 4 - \|-2 - 6\| - \|1 - 5\| - 5$	9. $\|-9 + 8 - 5 - 4\| - \|-6 - 9\| - \|(-8)(2)\|$
2. $\|-9 - 11\| - (-2)^2 + \|-11 + 9\| - \|3 - 9\|$	10. $\|-1 - 6 + 5\| - 4 \|-2 - 4 + 3\| - \|-6\|$
3. $\|1 - 9 - 7 + 1\| - 2 \|8 - 9\| - 5 + 2 \|8 - 9\|$	11. $\|6 - 11\| - 1^{16} - \|6 - 7 + 2\| + \|7 - 16\|$
4. $\|2 - 5\| + 2^5 - 3 \|2 - 5 - 5 + 2\| - \|-4\|$	12. $-\|3 - 10\| - \|4 - 9\| + \|-5 - 3\| - \|-3\|$
5. $\|-3 - 4 - 1\| + \|5 - 6 + 2 - 3\| - \|4 - 7\|$	13. $3^4 - \|-3 - 3\| + \|3 - 6\| - 3^3 - \|-6\|$
6. $\|-4 + 8 - 4 - 8\| - \|-8 - 4\| - (4)(-8) - 4$	14. $\|-8 - 9 - 3\| - \|-4 + 11 - 2\| - 10$
7. $\|-8 + 9 - 5 - 2\| - \|-7 - 9\| - 2 \|-7 + 5\|$	15. $-\|-2 - 5\| + \|-3 - 4\| - \|-6 - 1\| + 1$
8. $5^2 - \|2 - 5\| - \|5 - 2\| + 2^5 - \|-5\| + (-2)$	16. $\|3 - 4\| + 4^3 - \|4 - 3 - 4 + 3\| - 32$

Absolute Value 2

Solve by mental math.

1. $|x| = 6$	17. $|x + 3| = 8$	33. $|9 - x| = |-2|$
2. $|x| = -6$	18. $|x + 1| = 7$	34. $|x + 4| = |-15|$
3. $|-6| = x$	19. $|x - 5| = -1$	35. $|4 + x| = |-6|$
4. $|6| = -x$	20. $|x - 1| = 10$	36. $|x - 1| + 5 = 10$
5. $|-x| = 6$	21. $|3 - x| = 4$	37. $|2x + 4| = |-15 + 3|$
6. $x = |-5|$	22. $|x + 5| = 8$	38. $|2x + 3| = |-17 + 4|$
7. $|x| = -5$	23. $|x - 4| = 8$	39. $|x + 7| = |-13|$
8. $|-x| = 5$	24. $|x - 5| = 17$	40. $|x + 7| + 5 = 16$
9. $|5| = -x$	25. $|x + 7| = 14$	41. $|x - 8| + 2 = 14$
10. $|5| = x$	26. $|9 + x| = 20$	42. $|x - 6| - 7 = 8$
11. $|x| = 5$	27. $|6 - x| = -2$	43. $|x + 3| - 4 = 11$
12. $|x| = 7$	28. $|x + 6| = 15$	44. $3x + 5 = |-19 + 5|$
13. $|x| = -7$	29. $|2 - x| = 11$	45. $|x + 7| - 7 = 7$
14. $x = |7|$	30. $|x + 2| = 9$	46. $|5x + 5| = |-11 - 4|$
15. $|-7| = x$	31. $|6 + x| = 12$	47. $|2x + 4| = |-13 + 3|$
16. $|7| = -x$	32. $|8 - x| = 13$	48. $7x - 5 = |-21 - 2|$

Absolute Value 3

Find the least value of x such that:	Find the greatest value of x such that:	Find the number of integers x that satisfy:	Find the sum of the integral solutions.
1. |x + 1| ≤ 6	17. |x − 2| ≤ 8	33. |x + 3| ≤ 10	49. |x + 2| < 5
2. |x − 8| ≤ 11	18. |x + 2| ≤ 6	34. |x − 1| ≤ 4	50. |x − 1| ≤ 4
3. |x − 4| ≤ 10	19. |2x + 9| ≤ 17	35. |4x + 1| ≤ 9	51. |x + 3| < 7
4. |x + 5| ≤ 12	20. |4x + 6| ≤ 13	36. |5x − 4| ≤ 14	52. |x − 4| ≤ 6
5. |x + 7| ≤ 8	21. |3x − 2| ≤ 20	37. |x − 6| ≤ 2	53. |2x + 2| < 8
6. |2x + 7| ≤ 11	22. |x + 7| ≤ 11	38. |6x − 1| ≤ 7	54. |x + 5| ≤ 5
7. |x + 9| ≤ 15	23. |2x − 9| ≤ 5	39. |x + 9| ≤ 11	55. |x − 6| < 10
8. |x − 6| ≤ 13	24. |5x − 2| ≤ 12	40. |2x − 4| ≤ 5	56. |3x + 1| < 11
9. |3x + 4| ≤ 20	25. |2x + 4| ≤ 16	41. |8x − 1| ≤ 3	57. |3x − 3| ≤ 9
10. |x − 9| ≤ 16	26. |7x − 3| ≤ 9	42. |x + 4| ≤ 12	58. |4x + 3| ≤ 15
11. |2x + 1| ≤ 15	27. |2x + 5| ≤ 14	43. |3x − 4| ≤ 11	59. |x − 4| < 5
12. |x + 2| ≤ 15	28. |11x − 5| ≤ 7	44. |x − 3| ≤ 8	60. |x + 1| < 3
13. |4x − 3| ≤ 23	29. |6x + 6| ≤ 19	45. |3x − 2| ≤ 5	61. |2x + 1| ≤ 9
14. |x − 3| ≤ 17	30. |8x − 9| ≤ 21	46. |x + 3| ≤ 7	62. |10x + 3| < 15
15. |5x + 4| ≤ 21	31. |9x + 8| ≤ 1	47. |2x − 1| ≤ 7	63. |4x + 6| ≤ 10
16. |6x − 5| ≤ 29	32. |2x − 3| ≤ 0	48. |x + 5| ≤ 6	64. |3x + 2| < 8

Absolute Value 4

Solve.

1. | 3x − 7 | = 20	9. | −6 − 11x | = 60
2. 2 | 8 + 5x | = 66	10. −5 | 10x − 8 | = −75
3. 5 | 9x − 1 | = −10	11. 6 | 13 − 5x | = 0
4. | 7x + 9 | = 30	12. | 20x − 9 | = 51
5. | −4x − 12 | = 48	13. | 11x − 2 | = 20
6. 6 | −5x + 8 | = 72	14. − | 10x + 11 | = −39
7. | 14x + 7 | = 49	15. | 12x + 5 | = 51
8. −7 | 8x − 5 | = −84	16. 8 | 2x − 13 | = 72

Age Problems 1

Answer by making a vertical timeline. Distinguish between the relative (write on the left) and absolute (write on the right) ages.

1.	Freddie is 10 years older than Frenchie who is 6 years younger than Frannie. If Frannie is 17 now, how old is Freddie?	5.	Barry is 5 years older than Cary who is 3 years older than Harry. Mary, now 24, is 6 years younger than Barry. Find the sum of their ages.
2.	Katie is 6 years older than Kathy who is 2 years older than Cathy. Katy, who is 19, is five years older than Cathy. How old is Katie?	6.	Joe is 9 years older than Jo who is 4 years younger than Joan. If Josie is the oldest by 5 years and is now 22, how old is Jo?
3.	Shana is 9 years older than Jana who is 5 years younger than Dana. If Lana is the oldest by one year and is 24 now, how old is Jana?	7.	Jan is 20 years older than Jim who is 4 years younger than Jon. Jen, who is 52, is 4 years older than Jon. How old is Jan?
4.	TJ is 2 years older than PJ, and PJ is 6 years younger than RJ. If RJ is 20, how old are the other boys?	8.	Hank is younger than Hal by 6 years but older than Henry by 7 years. If Hank is now 20, find the sum of their ages.

Age Problems 2

Answer by making a chart.

1. Three years ago Shauna was one year younger than Liz will be in 4 years. If Shauna is 18 now, how old is Liz?

2. Hal's age in 30 years will be half a century. Six years ago, Bo was one year younger than Hal is now. How old will Bo be in 5 years?

3. Sarah's age is the average of Paul and Matt's. If Paul was 8 two years ago and Matt will be 20 in six years, how old is Sarah now?

4. Five years ago Karen was the same age as Mike now. Marc is now twice Mike's age. If Karen will be 18 in 3 years, how old was Marc 5 years ago?

5. Winston 2 years ago was as old as James is now. If Winston is now 16, how old will James be in five years?

6. Eli is 11 years old. When Eli is 23, the sum of his age and his brother's will be 53. How old is his brother now?

7. PJ's age is one-fifth greater than Ian's. Four years ago Ian's age was the least two-digit prime number. How old will PJ be in 7 years?

8. On the first day of 6th grade, Joy was one year younger than Vicki. Four years later Vicki was 16. How old was Joy when she started 2nd grade?

Angles 1

Find the complement of the angle in degrees by mental math.		Find the supplement of the angle in degrees by mental math.	
1. 70°	17. 64°	33. 130°	49. 70°
2. 8°	18. 11°	34. 46°	50. 31°
3. 35°	19. 66°	35. y°	51. 87°
4. 29°	20. 42°	36. 51°	52. 5g°
5. 89.5°	21. 29.8°	37. 92°	53. 111°
6. 60°	22. 14.7°	38. 99.5°	54. 14.7°
7. x°	23. m°	39. 73°	55. 154.6°
8. 23.4°	24. 61°	40. 24°	56. 9x°
9. 17°	25. 75.5°	41. 152°	57. 53°
10. 5°	26. 22°	42. 3w°	58. 178.3°
11. 13°	27. 41°	43. 106°	59. 1.5°
12. 54.9°	28. 39.6°	44. 43°	60. 99.3°
13. 72°	29. 3y°	45. 145°	61. 4h°
14. 54°	30. 21.5°	46. 68°	62. 138°
15. 33°	31. 50°	47. 19°	63. 103°
16. 2s°	32. 30.1°	48. 54.5°	64. 16.9°

Angles 2

Find the angle measure in degrees.

Answer as indicated.

1. Find the supplement of the complement of 25º.

2. Find the complement of the supplement of 127º.

3. Find the complement of the supplement of 110º.

4. Find the supplement of the complement of 42º.

5. Find the complement of the supplement of 154º.

6. Find the supplement of the complement of 69º.

7. Find the supplement of the complement of 51º.

8. Find the complement of the supplement of 163º.

9. Find the complement of the supplement of 145º.

10. Find the supplement of the complement of 78º.

11. Find the complement of the supplement of 172º.

12. Two supplementary angles are in the ratio 7:2. Find their product.

13. Two supplementary angles are in the ratio 5:4. Find their positive difference.

14. Two complementary angles are in the ratio 9:1. Find their positive difference.

15. Two complementary angles are in the ratio 2:1. Find their positive difference.

16. Two supplementary angles are in the ratio 7:5. Find their positive difference.

17. Two supplementary angles are in the ratio 8:7. Find the square of their difference.

18. Two complementary angles are in the ratio 3:7. Find their positive difference.

19. Two complementary angles are in the ratio 5:1. Find their positive difference.

20. Two complementary angles are in the ratio 11:7. Find their product.

21. Two supplementary angles are in the ratio 8:1. Find their product.

22. Two complementary angles are in the ratio 17:13. Find the supplement of the greater angle.

Angles 3

Label the angles in the diagram alternate exterior (AE), alternate interior (AI), corresponding (C), supplementary (S), or vertical (V).

1. ∠3 and ∠6	14. ∠5 and ∠8	27. ∠2, ∠3, and ∠4
2. ∠1 and ∠4	15. ∠11 and ∠12	28. (∠5 & ∠6) and ∠14
3. ∠12 and ∠13	16. ∠13 and ∠14	29. ∠1, ∠2, and ∠6
4. ∠2 and ∠8	17. ∠1 and ∠11	30. (∠3 & ∠4) and ∠9
5. ∠11 and ∠14	18. ∠2 and ∠10	31. (∠1 & ∠6) and ∠9
6. ∠2 and ∠5	19. ∠1 and ∠13	32. ∠1, ∠5, and ∠6
7. ∠4 and ∠13	20. ∠7 and ∠10	33. (∠2 & ∠3) and ∠14
8. ∠8 and ∠10	21. ∠7 and ∠9	34. (∠1 & ∠6) and ∠7
9. ∠5 and ∠10	22. ∠9 and ∠10	35. ∠1, ∠2, and ∠3
10. ∠4 and ∠11	23. ∠3, ∠4, and ∠5	36. (∠5 & ∠6) and ∠12
11. ∠12 and ∠14	24. (∠2 & ∠3) and ∠12	37. (∠2 & ∠3) and (∠5 & ∠6)
12. ∠11 and ∠13	25. ∠8 and ∠9	38. (∠3 & ∠4) and ∠7
13. ∠7 and ∠8	26. ∠4, ∠5, and ∠6	39. (∠1 & ∠2) and (∠4 & ∠5)

Angles 4

Solve algebraically.	*Answer as indicated. NTS*
1. What are the measures of two supplementary angles, the greater of which measures 4 times the lesser?	8. Find x + y. $y°$ $3y°$ $x°$
2. The sum of the measures of the complement and supplement of an angle is 230°. Find the angle.	9. Find x. $x°$ $135°$
3. The complement of an angle is 36° greater than twice the angle. Find the supplement of the angle.	10. Find y. $y°$ $x°$ $2x°$ $3x - 35°$
4. The sum of triple the complement of an angle and twice its supplement is 455°. Find the angle.	11. Find x. $x°$ $5y + 6°$ $y°$
5. The sum of one-half of the supplement of an angle and the complement of the angle is 135°. Find the supplement.	12. Find y. $x°$ $4x - 10°$ $y°$
6. Find the greater of two supplementary angles if one is 30° more than twice the other.	13. y + z = 185° Find x. $x°$ $y°$ $110°$ $z°$
7. Find the measure of an angle whose supplement is 50° less than three times its complement.	14. Find x. $115°$ $x°$

Angles 5

Find the missing angle measures using the diagram. Each problem is independent. NTS

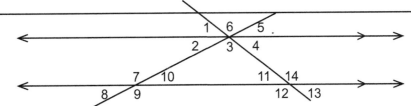

1. If m ∠ 3 = 42° and m ∠ 10 = 56°, then m ∠ 11 = _____ °.

2. If m ∠ 3 = 38° and m ∠ 8 = 45°, then m ∠ 11 = _____ °.

3. If m ∠ 11 = 25° and m ∠ 10 = 55°, then m ∠ 6 = _____ °.

4. If m ∠ 13 = 39°, then m ∠ 11 = _____ °.

5. If m ∠ 7 = 125° and m ∠ 1 = 40°, then m ∠ 6 = _____ °.

6. If m ∠ 11 = 26°, then m ∠ 4 = _____ °.

7. If m ∠ 7 = 110° and m ∠ 14 = 120°, then m ∠ 3 = _____ °.

8. If m ∠ 12 = 106° and m ∠ 7 = 99°, then m ∠ 3 = _____ °.

9. If m ∠ 5 = 65°, then m ∠ 10 = _____ °.

10. If m ∠ 5 = 62° and m ∠ 12 = 116°, then m ∠ 6 = _____ °.

11. If m ∠ 1 = 35° and m ∠ 3 = 55°, then m ∠ 5 = _____ °.

12. If m ∠ 8 = 42° and m ∠ 1 = 33°, then m ∠ 6 = _____ °.

13. If m ∠ 9 = 108° and m ∠ 6 = 44°, then m ∠ 13 = _____ °.

14. If m ∠ 8 = 44° and m ∠ 13 = 63°, then m ∠ 6 = _____ °.

15. If m ∠ 13 = 81° and m ∠ 6 = 36°, then m ∠ 2 = _____ °.

16. If m ∠ 13 = 76° and m ∠ 3 = 41°, then m ∠ 9 = _____ °.

17. If m ∠ 9 = 111° and m ∠ 6 = 22°, then m ∠ 14 = _____ °.

18. If m ∠ 9 = 103° and m ∠ 1 = 35°, then m ∠ 3 = _____ °.

Angles 6

Find x in degrees in the crook diagram. Assume parallel lines. NTS

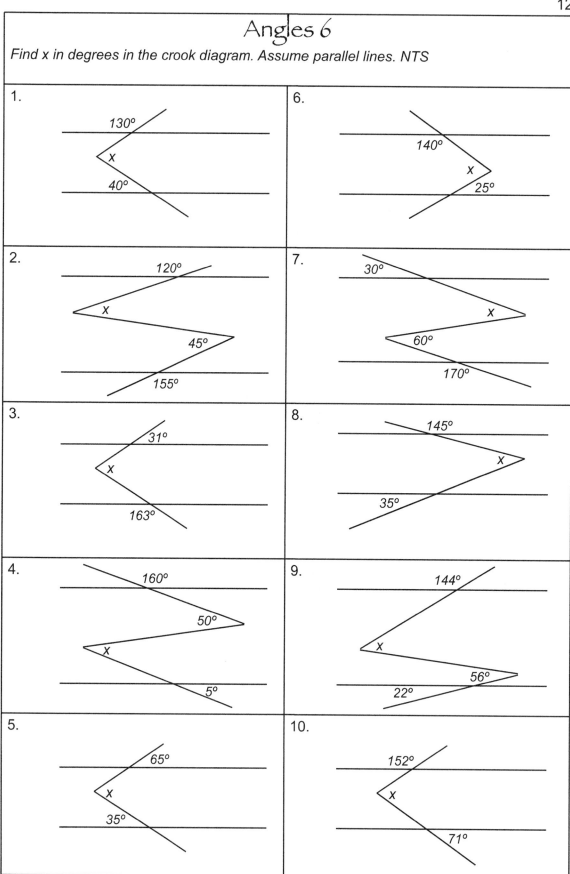

1.

130°

x

40°

6.

140°

x

25°

2.

120°

x

45°

155°

7.

30°

x

60°

170°

3.

31°

x

163°

8.

145°

x

35°

4.

160°

50°

x

5°

9.

144°

x

56°

22°

5.

65°

x

35°

10.

152°

x

71°

Area 1

Find the area in square units. Assume semicircles. One box equals one unit.

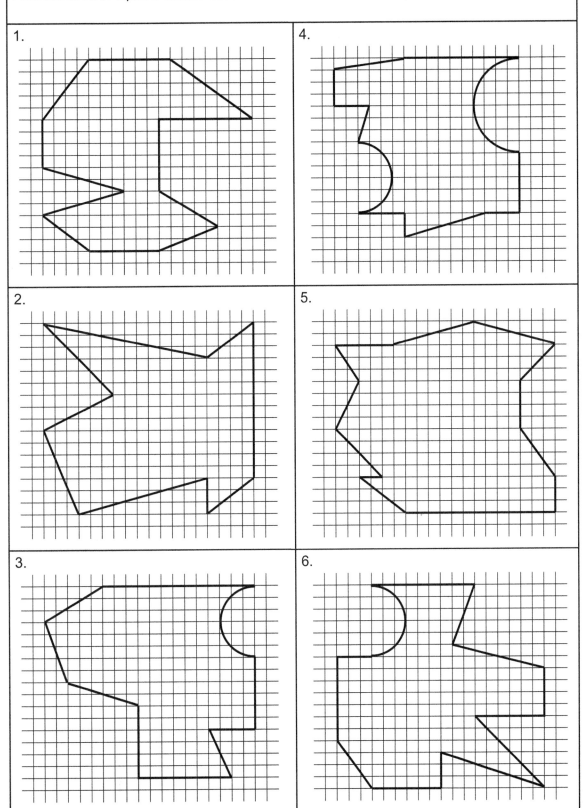

1.

2.

3.

4.

5.

6.

Area 2

Find the shaded area in square units. Assume right angles and tangency. Answer in terms of π. NTS

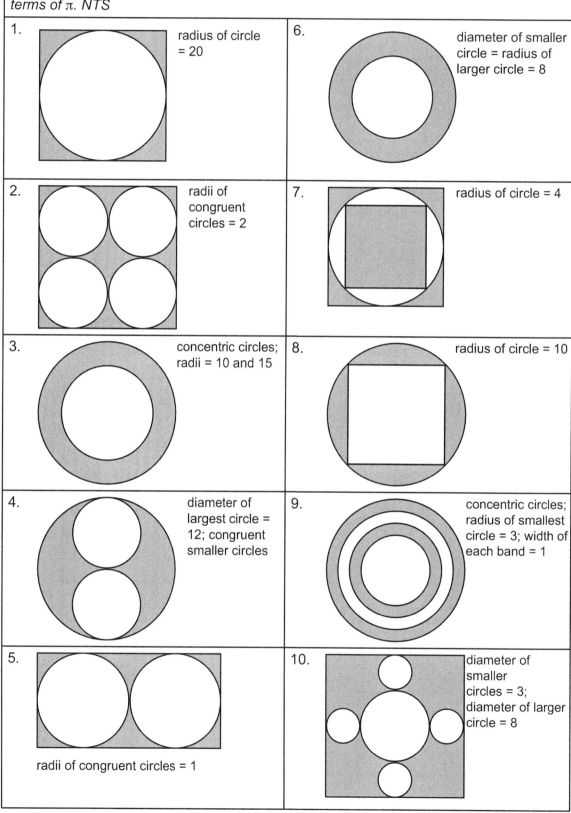

1. radius of circle = 20

2. radii of congruent circles = 2

3. concentric circles; radii = 10 and 15

4. diameter of largest circle = 12; congruent smaller circles

5. radii of congruent circles = 1

6. diameter of smaller circle = radius of larger circle = 8

7. radius of circle = 4

8. radius of circle = 10

9. concentric circles; radius of smallest circle = 3; width of each band = 1

10. diameter of smaller circles = 3; diameter of larger circle = 8

Area 3

Answer as indicated. All structures (photos, ponds, etc.) are rectangles. All borders are uniform width. All linear dimensions are units, and areas are square units.

1. A picture 7 by 11 has a border of size 3 all around. Find the area of the border.	6. A photo 5 longer than it is wide has a border of area 100 and width 2. Find the dimensions of the photo.
2. A photo 6 by 12 has a border of size 1.5 all around. Find the ratio of areas, border to photo.	7. A pool twice as long as it is wide has a border of area 940 and width 5. Find the dimensions of the pool.
3. A pond 12 by 15 has a border of size 2.5 all around. Find the ratio of areas, pond to border.	8. A garden 6 wider than it is long has a border of area 156 and width 3. Find the dimensions of the garden.
4. A pool 25 by 40 has a walkway all around. Find the width of the walkway if the total area is 1750.	9. A picture 4 longer than it is wide has a border of area 200 and width 2. Find the dimensions of the picture.
5. A garden 18 by 20 has a border of size 6 all around. Find the ratio of areas, garden to border.	10. A lawn with area 192 is triple as wide as it is long. Find the width of the lawn's border if the ratio of the areas, border to total, is 5:8.

Area 4

Find the area in square units of the rectangle bounded by:	Find the area in square units of the triangle bounded by both axes and the line:
1. $y = -3$, $x = -1$, $x = 9$, and $y = -11$	9. $y = -x + 4$
2. $x = 4$, $x = -5$, $y = 9$, and $y = -6$	10. $y = -2x - 6$
3. $x = -6$, $y = -2$, $x = 5$, and $y = -15$	11. $y = 2x - 8$
4. $y = -2$, $y = 13$, $x = -17$, and $x = -11$	12. $y = 3x + 3$
5. $x = 9$, $x = -3$, $y = 11$, and $y = -2$	13. $y = -x - 9$
6. $x = -12$, $y = 8$, $x = -2$, and $y = -11$	14. $y = 4x + 8$
7. $y = 1$, $y = -15$, $x = -19$, and $x = 11$	15. $y = 2x - 5$
8. $y = 6$, $x = -10$, $x = 4$, and $y = -8$	16. $y = -6x - 12$

Area 5

Find the area in square units using Pick's Formula: A = (B/2) + I − 1.
B = # border points; I = # interior points.

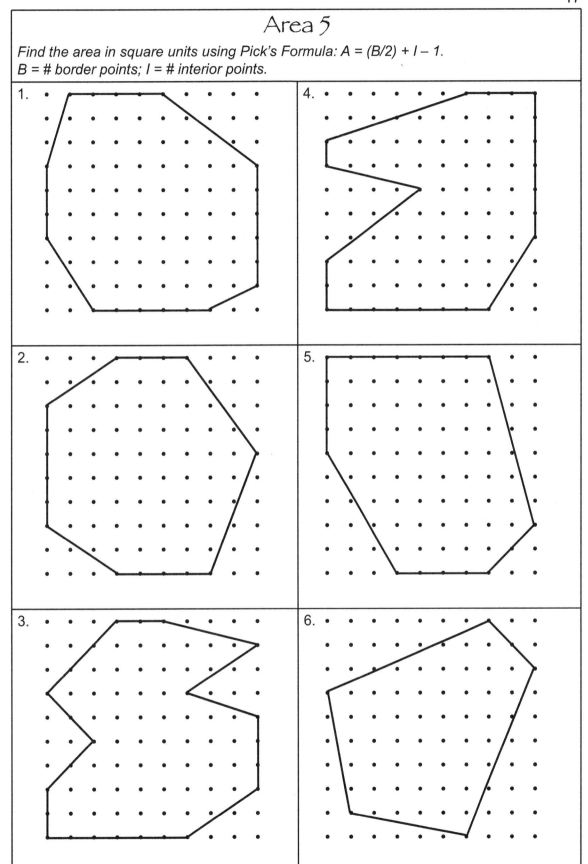

Area 6

Find the specified area in square units.

1. Find the area of the parallelogram bounded by x = 0, x = 4, y = 2x – 4, and y = 2x + 5.

2. Find the area of the parallelogram bounded by y = 5, y = –6, y = 3x – 5, and y = 3x + 7.

3. Find the area of the parallelogram bounded by y = 5, y = x, y = –4, and y = x + 5.

4. Find the area of the parallelogram bounded by x = –2, x = 3, y = 2x – 2, and y = 2x + 3.

5. Find the area of the trapezoid bounded by x = 6, y = 4, y = 2x, and the x-axis.

6. Find the area of the trapezoid bounded by x = 6, y = 8, y = –2x + 3, and x = –2.

7. Find the area of the trapezoid bounded by y = 3, y = –3, y = 4x – 5, and y = –4x – 5.

8. Find the area of the trapezoid bounded by x = 4, x = –2, y = (1/2)x – 4, and y = –x + 4.

Averages 1

Find the average by mental math. When summing, group to make multiples of 10.

1. 5, 825, 75, 95, 105, 155	12. 45, 15, 75, 95, 65, 45, 55, 85
2. 99, 53, 101, 50, 147	13. 32.4, 11.8, 88, 48.6, 19.2
3. 56, 420, 91, 60, 39, 60, 44	14. 161, 87, 113, 158, 142, 39, 70
4. 30, 19, 19, 11, 31, 51, 89, 70	15. 13.7, 12.1, 38.9, 37.3, 15.5, 2.5
5. 75, 76, 75, 76, 48	16. 40, 30, 20, 10, 60, 70, 80, 90
6. 56, 34, 72, 38, 91, 29, 170	17. 14, 99, 86, 11, 105, 75, 30
7. 109, 850, 391, 200, 150, 100	18. 38, 99, 51, 77, 23, 72
8. 100, 2000, 450, 950, 300, 400	19. 14, 19, 96, 21, 88, 32, 80
9. 820, 60, 20, 900, 200	20. 115, 698, 173, 227, 302, 285
10. 13, 87, 54, 46, 72, 28	21. 50, 75, 325, 150, 450, 225, 125
11. 90, 125, 300, 110, 375, 800	22. 17.8, 28, 49.2, 43.5, 22.5, 61

Averages 2

Find the average using the arithmetic sequence method.

1. 12, 14, 16, 18, 20	12. 102, 104, 106, 108, 110, 112
2. 1, 7, 13, 19, 25, 31, 37, 43	13. 1.4, 3.2, 5.0, 6.8
3. 91, 88, 85, 97, 94	14. 25, 32, 39, 46, 53, . . . , 81
4. 12, 15, 18, 21, . . . , 36	15. 84, 92, 86, 90, 82, 88
5. 4, 9, 14, 19, 24, 29	16. 979, 981, 983, 985, 987, 989
6. 1, 3, 5, 7, 9, 11, 13, 15, 17, 19	17. 3.5, 6.5, 9.5, 12.5, 15.5, 18.5
7. 117, 128, 139, 150	18. 17, 43, 69, 82, 30, 56
8. 20, 22, 24, 26, . . . , 38	19. 150, 300, 50, 200, 100, 350, 250
9. 26, 10, 6, 22, 18, 14	20. 4, 9, 14, . . . , 44
10. 20, 5, 25, 15, 10, 30	21. 40, 90, 60, 50, 30, 70, 80, 20
11. 29, 35, 41, 47, 53, . . . , 77	22. 6.1, 9.1, 12.1, 15.1, 18.1, 21.1

Averages 3

Find the average using the rightmost digits method.

1. 821, 826, 829, 827, 825, 822	12. 11, 17, 10, 14, 13, 18, 19, 17, 16
2. 621, 632, 602, 614	13. 34, 31, 35, 33, 33, 38
3. 91, 97, 94, 98, 95	14. 15, 19, 12, 13, 17
4. 738, 738, 738, 738, 738, 750	15. 3401, 3407, 3409, 3410, 3405
5. 29,325; 29,331; 29,326; 29,322	16. 103, 108, 107, 102, 110
6. 1202, 1209, 1217, 1211	17. 5407, 5406, 5411, 5404
7. 780, 780, 786, 784, 785	18. 1363, 1369, 1357, 1373, 1367, 1361
8. 80, 80, 81, 80	19. 731, 735, 739, 731
9. 521, 567, 555, 525	20. 1212, 1234, 1240, 1251, 1231
10. 29, 25, 24, 26	21. 414, 412, 419, 411
11. 818, 820, 819, 815, 825, 835	22. 327, 335, 322, 321, 320

Averages 4

Find the sum of the numbers.

1. Four numbers average to 14.	17. Six numbers average to 25.
2. Nine numbers average to 8.	18. Thirty numbers average to 13.
3. Twenty numbers average to 35.	19. Thirty numbers average to 80.
4. Twelve numbers average to 12.	20. Eleven numbers average to 75.
5. Forty numbers average to 15.	21. Eight numbers average to 61.
6. Eight numbers average to 70.	22. Fifty numbers average to 12.
7. Twenty-five numbers average to 40.	23. Eleven numbers average to 60.
8. Seven numbers average to 12.	24. Five numbers average to 71.
9. Five numbers average to 41.	25. Twenty numbers average to 62.
10. Fifty numbers average to 31.	26. Seven numbers average to 111.
11. Nine numbers average to 81.	27. Twelve numbers average to 110.
12. Eleven numbers average to 9.	28. Fifteen numbers average to 40.
13. Four numbers average to 70.	29. Nine numbers average to 31.
14. Eighty numbers average to 51.	30. Twenty numbers average to 41.
15. Thirty numbers average to 32.	31. Sixteen numbers average to 100.
16. Fifteen numbers average to 11.	32. Ten numbers average to 450.

Averages 5

Answer as indicated using the umbrella method.

1. Three numbers average to 26. If two of the numbers are 20 and 22, find the third number.

2. The average of four test scores is 78. If three of the scores are 56, 70, and 92, find the fourth score.

3. Three numbers average to 13. If two of the numbers are 4 and 23, find the third number.

4. The average of 3 numbers is 67. Two of the numbers are 56 and 91. Find the 3rd number.

5. If five test grades are 55, 56, 57, 58, and 59, what score on another test would raise the average to 60?

6. Two tests average to 85. What score on a third test would raise the average to 90?

7. Four test grades average to 78. What score on the next test would raise the average to 80?

8. The average of four test scores is 82. Three of the scores are 70, 85, and 90. Find the fourth score.

9. Four numbers average to 85. If three of the numbers are 70, 72, and 88, find the fourth number.

10. Test grades are 58, 60, 55, 50, and 55. What grade on the next test raises the average to 60?

Averages 6

Answer as indicated.

1. The average of 5 numbers is 85. Eliminating the least, the average is 91. Find the least number.

2. The average of 8 numbers is 12. Subtract 3 from 2 of the numbers. What is the new average?

3. The average of 7 numbers is 105. Eliminating the least, the average is 120. Find the least number.

4. The average of 5 tests is 80. If the lowest grade is dropped, 86 is the new average. Find the lowest grade.

5. The average of 9 numbers is 44. Subtract 6 from 3 of the numbers. What is the new average?

6. The average of 6 numbers is 312. Eliminating the greatest, the average is 280. Find the greatest number.

7. The average of 10 numbers is 84. After removing the 2 greatest and 2 least, the average is 72. Find the average of those removed.

8. The average of 7 tests is 68. After dropping the 2 lowest grades, the average is 78. Find the average of the 2 grades dropped.

9. The average of 4 test scores was 90. After removing two identical scores, the average was 88. What was the score removed twice?

10. The average of 20 numbers is 90. By removing the 3 greatest and 3 least, the average becomes 60. Find the average of those removed.

11. The average of 6 tests (range 0–100) is 70. If the first 4 tests average to 60, what is the lowest possible score on the last 2 tests?

12. The average of 8 tests (range 0–100) is 72. If the first 6 tests average to 64, what is the lowest possible score on the last 2 tests?

Bases 1

Count by writing the specified numerals.

1.	zero to ten in base four	12.	thirty-two to forty in base six
2.	three to thirteen in base five	13.	twenty-two to thirty in base twelve
3.	five to fifteen in base seven	14.	four to eleven in base two
4.	ten to nineteen in base six	15.	fifty-two to fifty-nine in base seven
5.	fifteen to twenty-two in base eight	16.	twenty to twenty-six in base five
6.	twelve to twenty-one in base sixteen	17.	thirty to thirty-eight in base eleven
7.	eight to nineteen in base eleven	18.	ten to eighteen in base three
8.	twelve to seventeen in base two	19.	eleven to nineteen in base four
9.	eight to nineteen in base nine	20.	thirty to forty in base sixteen
10.	nine to twenty in base twelve	21.	sixty-two to seventy in base eight
11.	three to twelve in base three	22.	fifty-five to sixty-four in base nine

Bases 2

Write the numeral one before.

1. E_{twelve}

2. 4000_{eight}

3. $13_{thirteen}$

4. $39C_{sixteen}$

5. 101100_{two}

6. $3T0_{eleven}$

7. 333_{four}

8. 560_{seven}

9. 1022_{three}

10. 800_{nine}

11. 20000_{four}

12. $635A_{fourteen}$

13. 1300_{five}

14. ET_{twelve}

15. 310_{eight}

16. $9TT_{eleven}$

Write the numeral one after.

17. 3133_{four}

18. 589_{eleven}

19. 366_{seven}

20. 199_{twelve}

21. $13_{thirteen}$

22. $1E_{twelve}$

23. 555_{six}

24. 477_{eight}

25. TTT_{eleven}

26. $2B5B_{sixteen}$

27. 212_{three}

28. 10111_{two}

29. $43EF_{sixteen}$

30. $7FF_{sixteen}$

31. $E_{fifteen}$

32. $5CD_{fourteen}$

Bases 3

Convert to base ten.

1. 343_{five}	8. 354_{seven}
2. 2385_{nine}	9. 1011011_{two}
3. $A2B_{sixteen}$	10. $T3E_{twelve}$
4. 22122_{three}	11. $3T8_{eleven}$
5. 562_{eight}	12. 2451_{six}
6. $13A_{fifteen}$	13. $25B_{thirteen}$
7. 2032_{four}	14. $1AC_{fourteen}$

Bases 4

Convert from base ten to the specified base.

1. $190 = ?_{five}$

8. $762 = ?_{four}$

2. $668 = ?_{eight}$

9. $279 = ?_{eleven}$

3. $94 = ?_{three}$

10. $1326 = ?_{twelve}$

4. $600 = ?_{seven}$

11. $1043 = ?_{fourteen}$

5. $696 = ?_{nine}$

12. $267 = ?_{six}$

6. $1234 = ?_{five}$

13. $729 = ?_{fifteen}$

7. $41 = ?_{two}$

14. $1952 = ?_{sixteen}$

Bases 5

Find the missing base b.

1. $102_b = 291_{ten}$	12. $101_b = 145_{ten}$
2. $11111_b = 31_{ten}$	13. $403_b = 103_{ten}$
3. $222_b = 114_{ten}$	14. $111_b = 13_{ten}$
4. $117_b = 79_{ten}$	15. $505_b = 325_{ten}$
5. $204_b = 166_{ten}$	16. $1001_b = 126_{ten}$
6. $105_b = 41_{ten}$	17. $1005_b = 1336_{ten}$
7. $103_b = 19_{ten}$	18. $111_b = 133_{ten}$
8. $907_b = 1096_{ten}$	19. $22_b = 34_{ten}$
9. $333_b = 63_{ten}$	20. $10002_b = 83_{ten}$
10. $402_b = 786_{ten}$	21. $1004_b = 220_{ten}$
11. $1003_b = 732_{ten}$	22. $111_b = 31_{ten}$

Bases 6

Convert among bases two, eight, and sixteen.

1. $111011_{two} = ?_{eight}$

12. $111011_{two} = ?_{sixteen}$

2. $777_{eight} = ?_{two}$

13. $10000001_{two} = ?_{sixteen}$

3. $567_{eight} = ?_{sixteen}$

14. $FF_{sixteen} = ?_{two}$

4. $314_{eight} = ?_{sixteen}$

15. $4AB_{sixteen} = ?_{eight}$

5. $1234_{sixteen} = ?_{eight}$

16. $C32_{sixteen} = ?_{eight}$

6. $5732_{eight} = ?_{sixteen}$

17. $4521_{eight} = ?_{sixteen}$

7. $1011101101_{two} = ?_{sixteen}$

18. $D7A_{sixteen} = ?_{eight}$

8. $1010101_{two} = ?_{eight}$

19. $10101011_{two} = ?_{eight}$

9. $1010101_{two} = ?_{sixteen}$

20. $10101011_{two} = ?_{sixteen}$

10. $ABC_{sixteen} = ?_{two}$

21. $2176_{eight} = ?_{two}$

11. $9DC_{sixteen} = ?_{eight}$

22. $EA1_{sixteen} = ?_{two}$

Bases 7

Operate in the given base.

1. 2135_{six} $+ 1424_{six}$	8. 5273_{eight} $+ 4365_{eight}$	15. 21012_{three} $+ 22102_{three}$	22. 3462_{seven} $+ 2536_{seven}$
2. 34_{five} $x\ 12_{five}$	9. 22_{three} $x\ 22_{three}$	16. 133_{four} $x\ 23_{four}$	23. 101_{two} $x\ 11_{two}$
3. 8341_{nine} $- 5278_{nine}$	10. 8375_{eleven} $- 38T6_{eleven}$	17. 3254_{eight} $- 1657_{eight}$	24. 2210_{three} $- 1012_{three}$
4. 1001011_{two} $+\ \ \ 10111_{two}$	11. 7235_{twelve} $+ 8967_{twelve}$	18. 10332_{four} $+ 31023_{four}$	25. 12343_{five} $+ 23342_{five}$
5. 53_{seven} $x\ 34_{seven}$	12. 45_{six} $x\ 23_{six}$	19. 475_{nine} $x\ 23_{nine}$	26. 394_{eleven} $x\ 53_{eleven}$
6. 2131_{four} $- 1333_{four}$	13. 3245_{seven} $- 1266_{seven}$	20. $418B_{sixteen}$ $- 23AD_{sixteen}$	27. 7654_{eight} $- 2356_{eight}$
7. 4323_{five} $+ 1443_{five}$	14. 2465_{nine} $+ 8574_{nine}$	21. $56T9_{eleven}$ $+ 4T78_{eleven}$	28. 1543_{six} $+ 3354_{six}$

Bases 8

Convert between the bases by using the given as one less than 1 in the next place.	Convert between the bases without passing through base ten.
1. $22222_{three} = ?_{ten}$	9. $475_{nine} = ?_{three}$
2. $55_{six} = ?_{two}$	10. $149_{twenty\text{-}five} = ?_{five}$
3. $1111111_{two} = ?_{five}$	11. $320112_{four} = ?_{sixteen}$
4. $888_{nine} = ?_{ten}$	12. $9A52_{sixteen} = ?_{four}$
5. $666_{seven} = ?_{ten}$	13. $1211201_{three} = ?_{nine}$
6. $4444_{five} = ?_{ten}$	14. $2863_{nine} = ?_{three}$
7. $TTT_{eleven} = ?_{ten}$	15. $233211_{five} = ?_{twenty\text{-}five}$
8. $EE_{twelve} = ?_{five}$	16. $A98_{twenty\text{-}seven} = ?_{three}$

Boundary 1

Answer as indicated.

1. $-3 \le x \le 12$
$4 \le y \le 15$
Find the least value of xy.

8. $-3 \le x \le 15$
$-15 \le y \le -2$
Find the greatest value of x − y.

2. $-10 \le x \le 30$
$-5 \le y \le 15$
Find the greatest value of x − y.

9. $2 \le x \le 25$
$-5 \le y \le 30$
Find the least value of xy.

3. $-10 \le x \le 11$
$-50 \le y \le -5$
A = greatest value of x − y
B = least value of x + y
Find A − B.

10. $-11 \le x \le 11$
$-12 \le y \le -10$
A = greatest value of x − y
B = least value of xy
Find A − B.

4. $6 \le x \le 24$
$-11 \le y \le -2$
A = least value of x − y
B = greatest value of x + y
Find A + B.

11. $-20 \le x \le -6$
$-30 \le y \le -15$
A = greatest value of x − y
B = least value of x + y
Find A − B.

5. $-20 \le x \le -8$
$-15 \le y \le -6$
Find the greatest value of xy.

12. $-6 \le x \le 9$
$4 \le y \le 20$
Find the least value of xy.

6. $-25 \le x \le 20$
$-50 \le y \le -25$
A = greatest value of x + y
B = least value of y − x
Find A + B.

13. $-10 \le x \le -2$
$2 \le y \le 10$
A = greatest value of x + y
B = least value of x − y
Find AB.

7. $-30 \le x \le 70$
$-10 \le y \le 20$
A = least value of x + y
B = greatest value of x − y
Find AB.

14. $4 \le x \le 35$
$11 \le y \le 20$
A = greatest value of x − y
B = least value of x + y
Find B − A.

Boundary 2

Answer as indicated.

1. The perimeter of a rectangle is 30. The width is less than 12. Find the range of values for its length.	8. The measure of an obtuse angle is $2x - 30$. Find the range of values for x.
2. The perimeter of a rectangle is 50. The length is greater than 10. Find the range of values for its width.	9. The perimeter of a rectangle is 90. The length is greater than 15. Find the range of values for its width.
3. The measure of an obtuse angle is $3x - 15$. Find the range of values for x.	10. The area of a rectangle is 144. Find its least and greatest perimeter if all sides are whole.
4. The measure of an acute angle is $2x - 12$. Find the range of values for x.	11. The perimeter of a rectangle is 144. Find its least and greatest area if all sides are whole.
5. The area of a rectangle is 100. Find its least and greatest perimeter if all sides are whole.	12. The perimeter of a rectangle is 70. The width is less than 25. Find the range of values for its length.
6. The perimeter of a rectangle is 100. Find its least and greatest area if all sides are whole.	13. The measure of an acute angle is $4x - 6$. Find the range of values for x.
7. A triangle has two vertices on a circle and one vertex at the center. If the central angle is less than 30°, find the range of values for the other 2 angles of the isosceles triangle.	14. A triangle has two vertices on a circle and one vertex at the center. If the central angle is less than 56°, find the range of values for the other 2 angles of the isosceles triangle.

Calendar 1

Find the day of the week.

1. 358 days from today if today is Friday	11. the 1st of the month if the 27th is Friday
2. 285 days ago if today is Saturday	12. the 4th of the month if the 29th is Monday
3. 421 days ago if today is Wednesday	13. the 31st of the month if the 4th is Friday
4. 80 days from today if today is Monday	14. the 3rd of the month if the 30th is Tuesday
5. 86 days ago if today is Thursday	15. the 27th of the month if the 3rd is Thursday
6. 290 days from today if today is Friday	16. the 4th of the month if the 26th is Monday
7. 216 days from today if 3 days ago was Tuesday	17. the 28th of the month if the 1st is Sunday
8. 149 days before yesterday if tomorrow is Sunday	18. the 30th of the month if the 5th is Saturday
9. 496 days from tomorrow if yesterday was Thursday	19. the 2nd of the month if the 31st is Wednesday
10. 723 days from today if today is Friday	20. the 29th of the month if the 5th is Monday

Calendar 2

Find the ones digit of the product of the calendar numbers marked by dots.

Find the positive difference of two products: the • numbers and the ø numbers.

1.

Su	M	T	W	R	F	Sa
				•		
				•		15
		•				
						•

5.

Su	M	T	W	R	F	Sa
		ø		•		
		•		ø		

2.

Su	M	T	W	R	F	Sa
	•	•				
					•	
						•
24		•				

6.

Su	M	T	W	R	F	Sa
ø					•	
•					ø	

3.

Su	M	T	W	R	F	Sa
		•		•		•
	15				•	
•						•

7.

Su	M	T	W	R	F	Sa
ø						•
•						ø

4.

Su	M	T	W	R	F	Sa
		•	•			•
				17	•	
						•
•						

8.

Su	M	T	W	R	F	Sa
		ø				•
		•				ø

Calendar 3

Find the day of the week, observing leap years.

1. June 1, 2005 was a Wednesday.
   June 1, 2015 was a _____.

2. July 14, 2014 was a Monday.
   July 14, 2009 was a _____.

3. January 1, 1991 was a Tuesday.
   January 1, 2000 was a _____.

4. January 1, 1900 was a Monday.
   January 1, 1904 was a _____.

5. February 2, 1992 was a Sunday.
   February 2, 1998 was a _____.

6. March 9, 1998 was a Monday.
   March 9, 2005 was a _____.

7. November 30, 2004 was a Tuesday.
   November 30, 2014 was a _____.

8. January 4, 2005 was a Tuesday.
   January 4, 2007 was a _____.

9. May 9, 2005 was a Mondday.
   May 9, 2008 was a _____.

10. December 12, 2004 was a Sunday.
    December 12, 2009 was a _____.

11. April 11, 2008 was a Friday.
    April 11, 2014 was a _____.

12. February 1, 2009 was a Sunday.
    February 1, 2012 was a _____.

13. January 1, 2015 was a Thursday.
    January 1, 2006 was a _____.

14. July 4, 2014 was a Friday.
    July 4, 2006 was a _____.

Calendar 4

Answer as indicated.

1. What day is the 420,716th day from a Friday?

2. If this month is April, what will be the month 1995 months from today?

3. If a full moon occurs once every 28 days, at most how many full moons can occur in one calendar year?

4. A fictitious calendar year has 14 months with even numbered months having 30 days and odd numbered months having 32 days. Find the number of days in the year.

5. A month with 5 Thursdays could start on which days of the week?

6. A calendar year has 13 months with prime-numbered months having 30 days, composite-numbered months have 25 days, and the 1st month having 35 days. Find the number of days in the year.

7. Suppose a calendar with 363 days has 11 months, each with 11 weeks. 438 of these days is how many years-months-weeks?

8. Find the date of the first Monday in a month given the sum of the dates of all Mondays in the month is 80.

9. A fictitious calendar year has 56 weeks with 8 days per week. Find the absolute difference of the number of days of this calendar and a standard one if neither has a leap year.

10. In what month does the 311th day of the year occur?

11. Find the least possible sum of the dates of all Mondays in any month.

12. Find the greatest possible sum of the dates of all Mondays in any month.

Ciphers 1

Decipher the addition problem. Each letter is a different digit.

1.
```
    A
+  BB
  AYY
```

2.
```
  ABA
+ BBA
  EYY
```

3.
```
  CBA
+ CCA
  BCC
```

4.
```
  AAB
+ PAB
  EPP
```

5.
```
  CAB
+ CAB
  AAC
```

6.
```
  ABB
+ BCB
  EAA
```

7.
```
  BAB
+ CAB
  ACC
```

8.
```
  ABA
+ ABY
  YYEB
```

9.
```
  ABA
  ABA
+ ABA
  PAPA
```

10.
```
  CBA
  CBA
+ CBA
  AAAC
```

11.
```
  CBA
  CBA
+ CBA
  ACAP
```

12.
```
   ABB
+ KPPB
  AYYDP
```

13.
```
  ABBC
+ KPBB
  PSCPA
```

14.
```
  ABCD
+ APPD
  DBAPB
```

15.
```
  ABPB
+ PCCK
  AAAAB
```

16.
```
  ABCD
+ PBCD
  PSPAA
```

MAVA Math: Enhanced Skills Copyright © 2015 Marla Weiss

Ciphers 2

Decipher the multiplication problem. Each letter is a different nonzero digit. The letter O is not used. A letter cannot be a digit already appearing in the problem.

1.
```
    BB
x    B
   A9B
```

2.
```
    A2
x   7A
  6396
```

3.
```
    AB
x    B
   249
```

4.
```
    AB
x    B
   A01
```

5.
```
   BBB
x    B
  277B
```

6.
```
   10A
x    A
   A2A
```

7.
```
   1A1
x    A
   A4A
```

8.
```
   AAB
x    B
  CB5B
```

9.
```
   A2C
x    2
  CPP2
```

10.
```
   AB2
x    B
   972
```

11.
```
    A6
x   5A
  2A8A
```

12.
```
   AAA
x    3
  BAAC
```

13.
```
   BBB
x    2
  AAAC
```

14.
```
   ABC
x    2
  BCD2
```

Circles 1

Complete the chart for each circle: radius, diameter, circumference, and area.

	RAD	DIAM	CIRC	AREA		RAD	DIAM	CIRC	AREA
1.	5				20.	80			
2.		30			21.			11π	
3.			100π		22.	25			
4.			9π		23.				100
5.				9π	24.	30			
6.			36π		25.				144π
7.			16π		26.				400π
8.				16π	27.		5		
9.		120			28.			100	
10.			π		29.			7π	
11.				π	30.				36
12.			38π		31.	11			
13.		12			32.		13		
14.	70				33.	π			
15.				256π	34.				81π
16.			4π		35.		π		
17.		14			36.			25π	
18.	14				37.	17			
19.	10				38.			20	

Circles 2

Answer as indicated.

1. The ratio of the circumferences of two circles is 9:16. Find the ratio of their areas.

2. A circle with circumference 2 has the same area as a square. Find the perimeter of the square.

3. Points A, B, C, D, E, and F in order are on a circle of area 81π, dividing it into 6 congruent arcs. Find the length of the arc AEC.

4. Four concentric circles have radii 2, 4, 6, and 8. Find the sum of the areas of the outermost and innermost rings.

5. Find the area of the rectangle if the congruent circles with centers shown have diameter 10.

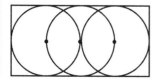

6. Three circles are pairwise externally tangent. The triangle formed by joining their centers has sides 8, 9, and 10. Find the three diameters.

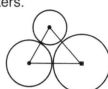

7. Quadrilateral ABCD is circumscribed about a circle. AB = 19. BC = 15. AD = 13. Find CD.

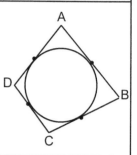

8. Congruent semicircles comprise the curved path of length 55π. Find the area of the rectangle as shown.

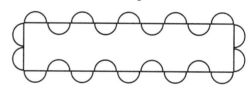

9. A circle with circumference 30 has two overlapping arcs of lengths 15 and 25. Find the positive difference of the greatest and least possible overlap of the arcs.

10. Find the area of circle O if the smaller square has area 11.

Circles 3

Calculate for the sector. CENTRAL ANGLE	FRACTION OF CIRCLE	RADIUS	AREA OF SECTOR	LENGTH OF ARC	PERIMETER OF SECTOR
1. 45°		4			
2. 90°		8			
3. 120°		12			
4. 80°		9			
5. 100°		18			
6. 180°		30			
7. 20°		36			
8. 60°		18			
9. 135°		32			
10. 150°		24			
11. 30°		24			
12. 50°		72			
13. 10°		36			
14. 160°		45			
15. 40°		27			
16. 30°		36			
17. 140°		54			
18. 240°		15			
19. 225°		32			

Circles 4

Answer as indicated.

1. $\overline{AD}$ is a diameter. Find the sum of the measures of the 4 angles of the triangles not opposite the diameter.

6. 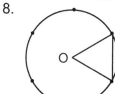 O is the center of the circle with radius 8. Find the area of the circle not inside the △.

2. 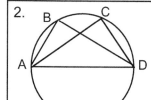 $\overline{AD}$ is a diameter. AD = 85. AB = 13. CD = 40. Find AC + BD.

7. O is the center of the circle with radius 6. Find the area of the circle not inside either △.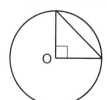

3. 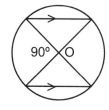 Each △ has one vertex at the center of the circle. Find the sum of the measures of the 4 noncentral angles of the △s.

8. 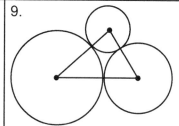 The 6 points are equally spaced on the circle with center O. Find the measures of the 3 angles of the △.

4. 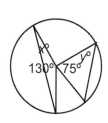 Each △ has one vertex at the center of the circle. Find x + y.

9. 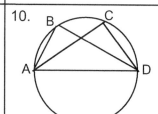 Each vertex of the △ is a center of a circle. Find the ratio of the sum of the 3 circumferences to the perimeter of the △.

5. 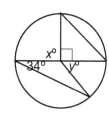 Each △ has one vertex at the center of the circle. Find x + y.

10. 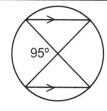 $\overline{AD}$ is a diameter. AD = 25. BD = 24. AC = 20. Find AB + CD.

Circles 5

Find the area in square units.

1. Find the area of a circle inscribed in a square of area 36.

2. Find the area of a circle inscribed in a square with side 5.

3. Find the area of a circle inscribed in a square of area 10.

4. Find the area of a square circumscribed about a circle with radius 7.5.

5. Find the area of a square circumscribed about a circle with circumference 12π.

6. Find the area of a square circumscribed about a circle with area 49π.

7. Find the area of a square inscribed in a circle with area 9π.

8. Find the area of a square inscribed in a circle with radius 2.

9. Find the area of a square inscribed in a circle with area 20π.

10. Find the area of a circle circumscribed about a square with area 1.

11. Find the area of a circle circumscribed about a square with side 5.

12. Find the area of a circle circumscribed about a square with area 10.

Circles 6

Answer for the inscribed polygon.	Find the length in units of the common external tangent of the two circles.
1. An equilateral triangle is inscribed in circle with radius 6. Find the perimeter of the triangle.	7. radii 2 and 10, centers 17 apart
2. A regular hexagon is inscribed in circle with circumference 18π. Find area of the hexagon.	8. radii 8 and 18, externally tangent to each other
3. A regular octagon is inscribed in circle. Find least interior angle of the triangle formed by connecting 3 adjacent vertices.	9. radii 3 and 14, centers 61 apart
4. A regular hexagon is inscribed in circle with circumference 16π. Find perimeter of the hexagon.	10. radii 3 and 10, centers 25 apart
5. A regular pentagon is inscribed in circle. Find least interior angle of the triangle formed by connecting 3 adjacent vertices.	11. radii 9 and 16, externally tangent to each other
6. An equilateral triangle is inscribed in circle with radius 8. Find the area of the triangle.	12. radii 2 and 8, externally tangent to each other

Clocks 1

Convert units.

1.	.5 day =	_____ hours	20.	1/3 minute =	_____ seconds
2.	5/6 hour =	_____ minutes	21.	.15 hour =	_____ minutes
3.	.25 day =	_____ hours	22.	1.1 minute =	_____ seconds
4.	.25 hour =	_____ minutes	23.	.4 hour =	_____ minutes
5.	.1 hour =	_____ minutes	24.	.75 hour =	_____ minutes
6.	.4 minutes =	_____ seconds	25.	2/3 hour =	_____ minutes
7.	1/3 hour =	_____ minutes	26.	.45 hour =	_____ minutes
8.	.85 hour =	_____ minutes	27.	.8 hour =	_____ minutes
9.	5/6 day =	_____ hours	28.	1/6 day =	_____ minutes
10.	.6 hour =	_____ minutes	29.	.9 hour =	_____ minutes
11.	1/12 day =	_____ minutes	30.	.25 year =	_____ months
12.	5/12 minute =	_____ seconds	31.	.7 hour =	_____ minutes
13.	.1 hour =	_____ seconds	32.	7/8 day =	_____ hours
14.	3/5 hour =	_____ minutes	33.	.15 minute =	_____ seconds
15.	.75 minute =	_____ seconds	34.	.05 hour =	_____ minutes
16.	.3 hour =	_____ minutes	35.	7/12 minute =	_____ seconds
17.	.125 day =	_____ hours	36.	7/15 hour =	_____ seconds
18.	.2 hour =	_____ minutes	37.	11/30 hour =	_____ minutes
19.	4/15 minute =	_____ seconds	38.	11/12 day =	_____ minutes

Clocks 2

Answer as indicated.

1. What time is 3540 seconds before 2:57 PM?

2. If Eli goes to sleep at 9:30 PM and sleeps for 10 hours and 30 minutes, when will he wake up?

3. What time is 10 hours and 58 minutes after 3:19 PM?

4. The time on a 12-hour clock is 10:00 AM. Find the time after the minute hand goes around 4 times.

5. The number of minutes in 5 hours is the same as the number of hours in how many days?

6. What time is 335 minutes after 2:55 PM?

7. Find the time 281 hours ago if the present time is 7:00 AM.

8. If a clock stopped running 313 minutes after 3:17 PM, at what time did it stop?

9. The time on a 12-hour clock is 9:00 AM. Find the time after the minute hand goes around 3 times.

10. What time is 2460 seconds before 4:39 PM?

11. What time is 8 hours and 52 minutes after 1:37 PM?

12. If a clock stopped running 257 minutes after 1:54 PM, at what time did it stop?

13. What time is 285 minutes after 4:25 PM?

14. On any day, 8:39 AM is how many minutes before 3:26 PM?

Clocks 3

Operate as indicated. Regroup/simplify answers.

1.   13 days   8 hours   9 min 55 sec
  + 5 days 16 hours 54 min   7 sec
	9.
	    7 days 7 hours 7 min 7 sec
	  − 5 days 7 hours 9 min 11 sec

2.
 	
    13 days 6 hours   9 min 3 sec
  − 5 days 16 hours 54 min 5 sec
	10.
	
	    18 days 20 hours 40 min 1 sec
	  − 10 days 22 hours 50 min 9 sec

3. (9) (5 days 7 hrs 7 min 7 sec)	11. $\frac{1}{4}$ (5 days 9 hours 8 minutes)
4. $\frac{1}{3}$ (7 days 8 hrs 9 min 42 sec)	12. (4) (2 days 7 hr 20 min 40 sec)
5. (10 hrs 22 min 40 sec) ÷ 4	13. (6 hrs 30 min 16 sec) ÷ 4

6.
 	
    9 hours
  − 5 hours 5 minutes 5 seconds
	14. 4 days 10 hours 22 min 32 sec
	  + 3 days 16 hours 49 min 31 sec

7. (11 hrs 31 min 20 sec) ÷ 5	15. $\frac{1}{6}$ (8 days 8 hours 6 minutes)
8. (13 hrs 16 min 18 sec) ÷ 6	16. (14 hrs 10 min 15 sec) ÷ 3

Clocks 4

Find the measure in degrees of the angle formed by the hands of a clock at the given time.

1. 12:30

2. 6:45

3. 1:20

4. 10:30

5. 3:15

6. 11:20

7. 5:40

8. 2:30

9. 8:10

10. 12:15

11. 6:40

12. 1:45

13. 4:45

14. 9:20

15. 3:05

16. 9:30

17. 2:45

18. 7:30

19. 8:15

20. 7:10

21. 3:20

Clocks 5

Find the fractional part.	*Find the time on the nonstandard clock, disregarding AM and PM.*	
1. One and one half hours are what fraction of the time between 12 PM on Monday and 12 AM on Saturday of the same week?	8. 8-hour clock start at 7:00 22 hours later	15. 4-hour clock start at 2:00 19 hours later
2. Two and one third hours are what fraction of the time between 12 PM on Tuesday and 12 AM on Friday of the same week?	9. 13-hour clock start at 11:00 56 hours later	16. 9-hour clock start at 1:00 70 hours later
3. Two hours and fifteen minutes are what fraction of the time between 8 PM on Wednesday and 11 AM on Friday of the same week?	10. 7-hour clock start at 4:00 17 hours later	17. 20-hour clock start at 15:00 87 hours later
4. Three and one third hours are what fraction of the time between noon on Wednesday and noon on Saturday of the same week?	11. 15-hour clock start at 14:00 68 hours later	18. 14-hour clock start at 12:00 30 hours later
5. One hour and fifty minutes are what fraction of the time between 10:00 AM Monday and 7:00 PM Tuesday of the same week?	12. 10-hour clock start at 9:00 35 hours later	19. 16-hour clock start at 12:00 39 hours later
6. Three hours and twelve minutes are what fraction of the time between noon Sunday and 4:00 PM Tuesday of the same week?	13. 6-hour clock start at 5:00 79 hours later	20. 17-hour clock start at 8:00 29 hours later
7. Four hours and ten minutes are what fraction of the time between 9:00 AM Sunday and 1:00 PM Thursday of the same week?	14. 11-hour clock start at 2:00 37 hours later	21. 5-hour clock start at 3:00 44 hours later

Clocks 6

Answer as indicated for the broken clock.

1. A clock that uniformly loses 5 minutes every 12 hours was correctly set at noon on January 1. Find the time on the clock when the correct time was noon on January 5 of the same year.

2. A clock is correctly set at 4:00 PM, but it loses 4 minutes every hour. What is the correct time when the clock reads 8:00 AM the next day?

3. A 12-hour clock loses 5 minutes each day. Find the number of days for the clock to first return to the correct time.

4. The hands on a broken 12-hour clock moved counterclockwise. Find the number to which the minute hand pointed 15 minutes before the hour hand pointed to 4.

5. Two clocks show the correct time at 2:00. One runs backward and the other forward, both at the normal rate. When will both clocks next show the same time?

6. A broken clock runs at the normal rate but backward. A 2nd broken clock runs forward at twice the normal rate. Both start at 12. Find the correct time when the clocks next show the same time.

7. A clock is correctly set at 2:00 AM, but it loses 5 minutes every hour. What is the correct time when the clock reads 9:00 PM the next day?

8. A 12-hour clock loses 15 minutes each day. Find the number of days for the clock to first return to the correct time.

9. The hands on a broken 12-hour clock moved counterclockwise. Find the number to which the minute hand pointed 20 minutes before the hour hand pointed to 5.

10. A clock that uniformly loses 6 minutes every 24 hours was set correctly at noon on May 1. Find the time on the clock when the correct time was 4:00 AM on May 6 of the same year.

Combinations 1

Calculate.

1. C(5, 2)	12. C(15, 14)
2. C(8, 3)	13. C(20, 2)
3. C(7, 4)	14. C(11, 8)
4. C(10, 5)	15. C(8, 6)
5. C(6, 2)	16. C(7, 5)
6. C(12, 4)	17. C(9, 5)
7. C(9, 6)	18. C(12, 9)
8. C(11, 9)	19. C(10, 7)
9. C(4, 3)	20. C(20, 3)
10. C(10, 6)	21. C(8, 5)
11. C(9, 7)	22. C(13, 11)

Combinations 2

Answer as indicated.	*Answer as indicated.*
1. Nineteen telephones may be paired for conversations in how many ways?	9. How many handshakes occur among 31 people if everyone shakes hands once?
2. How many ways can a doubles team be selected from 8 players?	10. How many gifts are exchanged among 10 people if everyone exchanges gifts with every other person?
3. How many ways can a club of 6 members choose a committee of 3 people?	11. How many rolls occur if each of 13 children sitting in a circle rolls a ball to each other child?
4. How many ways can 3 people be chosen from 4 couples if all are equally eligible?	12. How many cards are used if all 20 students in a class exchange valentines.
5. How many ways can a reader select 3 books from 5 different books?	13. Find the least number of line segments to connect 12 points on a circle.
6. How many pairs of friends may be chosen from among 6 friends?	14. In a league of 16 teams, how many games are played matching each team with every other team once?
7. How many triangles may be formed selecting 3 of 7 points on a circle as the vertices?	15. In a league of 14 teams, how many games are played matching each team with every other team twice?
8. How many quadrilaterals may be formed selecting 4 of 6 points on a circle as the vertices?	16. How many ways may 2 books be paired from a collection of 20 books?

Cones 1

Find the volume of the cone in cubic units. | *Find the radius of the cone in units.*

1. radius = 10, height = 12

9. volume = 30π, height = 10

2. radius = 6, slant height = 10

10. volume = 132π, height = 11

3. height = 10, slant height = 26

11. volume = 15π, height = 5

4. radius = 15, altitude = 7

12. volume = 90π, height = 10

5. diameter = 14, slant height = 25

13. volume = 135π, height = 5

6. diameter = 12, altitude = 10

14. volume = 576π, height = 12

7. height = 40, slant height = 41

15. volume = 72π, height = 6

8. height = 60, slant height = 61

16. volume = 80π, height = 12

Cones 2

Answer as indicated. Measurements are units, square units, and cubic units.

1. A cone "just fits" in a cylinder with diameter 10 and height 12. Find the volume between the two solids.

2. Two upright cones with heights 6 and 12 share a base with radius 8. Find the volume remaining when the smaller cone is removed from the larger cone.

3. A cylinder with height 10 has a hemisphere with equal circumference attached at one end and a cone with altitude 16 and slant height 20 attached at the other end. Find the total volume.

4. A cylinder with height 12 has a hemisphere attached at one end and a cone at the other, base to base. The cone has height 4 and slant height 5. Find the total surface area.

5. If the radius of a cone is tripled and the height is doubled, find the percent increase in the volume.

6. The height of a cone is reduced by 1/3, and the radius is increased by 1/2. The volume of the new cone is what fraction of the volume of the original cone?

7. Two upright cones with heights 3 and 9 share a base with radius 6. Find the volume remaining when the smaller cone is removed from the larger cone.

8. A sphere of ice cream has radius 2. If the ice cream sits on an ice cream cone with the same radius and melts into the cone without dripping, find the height of the cone to hold all of the ice cream.

Cones 3

A right triangle with one vertex at the origin is rotated about an axis. Find the volume in cubic units of the resulting cone with the given specifications.

1.   vertices = (12,0) and (12, 5)
      rotated about *x*-axis
	5. vertices = (0,10) and (6,10)
	      rotated about *y*-axis

2.   vertices = (18,–5) and (0, –5)
      rotated about *y*-axis
	6. legs = 9
	      rotated about *x*-axis

3.   one vertex = (15,0); one leg = 6
      rotated about *x*-axis
	7. legs = 30
	      rotated about *y*-axis

4.   one vertex = (0,9); hypotenuse 15
      rotated about *y*-axis
	8. vertices = (21,7) and (21, 0)
	      rotated about *x*-axis

Cones 4

Cut a circular paper into congruent sectors. Form one sector into a cone with no overlap. Find the volume of the cone given two facts.

1.   radius of circular paper = 9
      number of congruent sectors = 3
	5. radius of circular paper = 10
	      number of congruent sectors = 5

2.   radius of circular paper = 8
      number of congruent sectors = 4
	6. radius of circular paper = 40
	      number of congruent sectors = 10

3.   radius of circular paper = 18
      number of congruent sectors = 6
	7. radius of circular paper = 20
	      number of congruent sectors = 5

4.   radius of circular paper = 16
      number of congruent sectors = 8
	8. radius of circular paper = 24
	      number of congruent sectors = 6

Cones 5

Find the surface area of the cone in square units.

1. radius = 10, slant height = 12	10. altitude = 8, slant height = 10
2. radius = 6, slant height = 11	11. altitude = 24, slant height = 25
3. diameter = 10, slant height = 13	12. altitude = 30, slant height = 34
4. radius = 15, slant height = 7	13. altitude = 60, slant height = 61
5. diameter = 14, slant height = 25	14. altitude = 15, slant height = 17
6. diameter = 18, slant height = 11	15. altitude = 14, slant height = 50
7. radius = 20, slant height = 21	16. altitude = 21, slant height = 29
8. diameter = 16, slant height = 17	17. altitude = 9, slant height = 15
9. radius = 11, slant height = 18	18. altitude = 24, slant height = 26

Cones 6

Find the volume of the cone frustrum in cubic units.

1. upper frustrum base radius = 5
   lower frustrum base radius = 15
   frustrum height = 24

2. upper frustrum base radius = 9
   lower frustrum base radius = 12
   frustrum slant height = 5

3. larger cone height = 40
   smaller cone height = 20
   frustrum slant height = 29

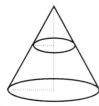

4. upper frustrum base radius = 6
   lower frustrum base radius = 9
   frustrum slant height = 5

5. larger cone height = 45
   smaller cone height = 15
   frustrum slant height = 34

6. larger cone base area = 36
   larger cone altitude = 18
   frustrum altitude = 6

7. upper cone height = 12
   frustrum height = 8
   frustrum slant height = 10

8. upper frustrum base area = 4
   lower frustrum base area = 16
   smaller cone altitude = 6

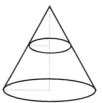

Coordinate Plane 1

Answer as indicated. All rectangles and squares are parallel to the axes. Linear measurements are units; area measurements are square units

1. Rectangle ABCD has vertices A(2,−3) and C(−1,4). Find vertices B and D.

8. Find the coordinates of the point equidistant from points A(7, 4), B(−1, 4), C(−1, −2), and D(7, −2).

2. Rectangle EFGH has vertices E(−5,8) and G(9,−7). Find vertices F and H.

9. ORST is a square in Quadrant II. Point O is the origin. If point S is (k, 4), what is the value of k?

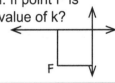

3. Find the area of a rectangle with two vertices (9,2) and (−1, −13).

10. OEFG is a square in Quadrant III. Point O is the origin. If point F is (−6, k), what is the value of k?

4. Find the area of a rectangle with two vertices (7,10) and (−4, −2).

11. A square with coordinates A(−6,−3), B(−6, 6), C(3, 6), and D(3,−3) is translated 3 units in the positive y direction and then reflected across the x-axis. Find the new coordinates of C.

5. Find the area and perimeter of the rectangle with vertices A(9,2), B(9,−6), C(−5, −6), and D(−5, 2).

12. Find the 2 missing coordinates of the rectangle.

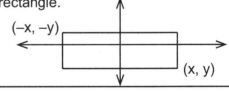

6. A square with coordinates A(2,−1), B(7, −1), C(7, 4), and D(2,4) is translated 4 units in the positive x direction and then reflected across the y-axis. Find the new coordinates of C.

13. Find the coordinates of the point equidistant from points A(9, 6), B(−2, 6), C(−2, −3), and D(9, −3).

7. OABC is a square in Quadrant I with coordinates O at the origin and A(3a, 0). Find the coordinates of B and C.

14. Find the area and perimeter of the rectangle with vertices A(8,7), B(8,−8), C(−7, −8), and D(−7, 7).

Coordinate Plane 2

Answer as indicated.

1. A circle with center at the origin passes through (0,7). Name 3 more points on the circle.

7. Name a point on an axis that is twice as far from (2,3) as it is from (1,2).

2. Points A (2,1), B(8,1), and C(x,y) are such that the area of △ABC is 21. Find the 2 possible values for y.

8. A circle with center (0,0) has a diameter with one endpoint (5,6). Find the other endpoint.

3. A rhombus with 2 sides parallel to the x-axis has vertices at (6, 15), (1, 3), and (14, 3). Find the 4th vertex.

9. Travel from E to C on rectangle ABCD. How much farther is the trip through A and D than the trip through B?

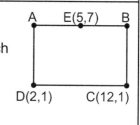

4. Find the coordinates of D such that ABCD is an isosceles trapezoid.

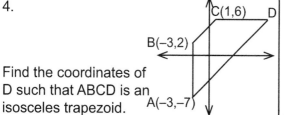

10. The area of the largest rectangle is 6 times the area of the smallest. Find the coordinates of A.

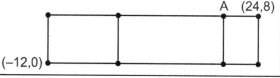

5. A circle with center at (−2, −5) passes through (−2, 3). Name 3 more points on the circle.

11. Points A (6,2), B(10,2), and C(x,y) are such that the area of △ABC is 12. Find the 2 possible values for y.

6. $\overline{AB}$ is a diameter of the circle with center at the origin. Find the coordinates of B.

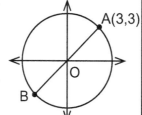

12. Find the area of the triangle with coordinates (1,0), (7,0), and (9,8).

Coordinate Plane 3

Find the area in square units of the convex or concave polygon determined by the coordinates.

1. (0,0), (0,4), (9,8), (18,8), (18,4), (9,0)

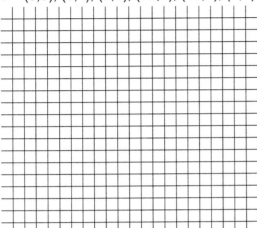

4. (0,1), (2,8), (8,4), (8,1), (2,−1)

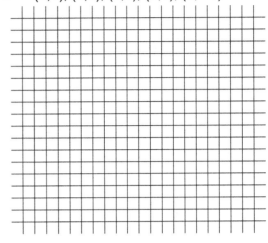

2. (−5,7), (−2,3), (6,−1)

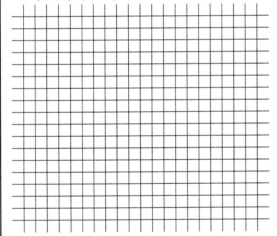

5. (8,−2), (4,−5), (1,3), (1,6), (10,8)

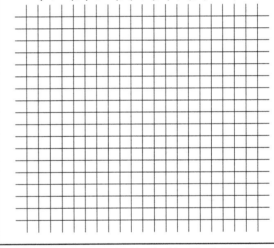

3. (9,2), (5,4), (2,8), (−6,3), (−3,−3)

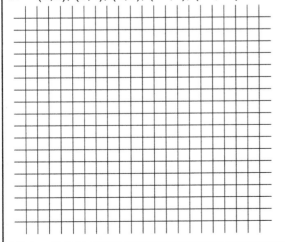

6. (−9,1), (−5,3), (6,1), (1,−7), (−7,−4)

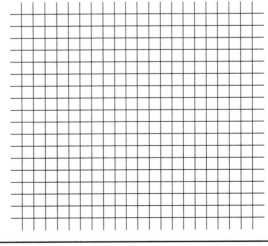

Find the reflection of each point across the:	x-axis.	y-axis.	line y = x.	line y = −x.	origin.
Coordinate Plane 4					
1. (1, 4)					
2. (2, 7)					
3. (−1, 5)					
4. (−7, −2)					
5. (0, 9)					
6. (8, 0)					
7. (6, −3)					
8. (5, 8)					
9. (−2, −1)					
10. (−3, 3)					
11. (4, −7)					
12. (3, −4)					
13. (9, 9)					
14. (10, 1)					
15. (11, 2)					
16. (−5, −6)					

Counting 1

Answer by drawing a Venn diagram.

1. Of 83 houses, 37 need new roofs and 46 need new shrubs. If 19 need both, how many need neither?

2. Of 98 students, 63 take math, 42 take music, and 18 take both math and music. How many students take neither?

3. Of 62 girls, 43 are wearing sweaters, 32 are wearing coats, and 16 are wearing both. How many are wearing neither?

4. Of 181 dresses, 73 have buttons, 102 have zippers, and 29 have both. How many have neither?

5. Of 100 students, 52 take Spanish, 73 take German, and 25 take both languages. How many students take neither?

6. Show the relationship among the factors of 18, 30, and 45.

7. Show the relationship among the factors of 20, 50, and 60.

8. Show the relationship among the factors of 24, 40, and 42.

9. Show the relationship among the factors of 16, 28, and 35.

10. Show the relationship among the infinite sets N, W, Z, Q, and R. Write the digit 0 and "irrationals" in the correct band.

Counting 2

Answer by drawing a Venn diagram.

1. A survey of 300 children found:
   45 drink juice, milk, and cocoa;
   60 drink juice and milk;
   70 drink milk and cocoa;
   46 drink juice and cocoa;
   160 drink milk;
   75 drink juice; and
   110 drink cocoa.
   How many drink none of the three?

2. A survey of 250 students found:
   36 study French, Spanish, and Latin;
   142 study French;
   86 study Spanish;
   108 study Latin;
   56 study French and Latin;
   76 study Latin and Spanish;
   42 study French and Spanish;
   How many study none of the three?

3. For sets A, B, and C:
   n(A) = 100
   n(B) = 80
   n(C) = 60
   n(A AND B) = 32
   n(B AND C) = 28
   n(A AND C) = 40
   n(A AND B AND C) = 12
   n(B′) = 74
   Find n(A OR B OR C)′.

4. For sets A, B, and C:
   n(A) = 109
   n(B) = 111
   n(C) = 153
   n(A AND B) = 51
   n(B AND C) = 63
   n(A AND C) = 39
   n(A AND B AND C) = 21
   n(A′) = 133
   Find n(A OR B OR C)′.

Counting 3

Answer by constructing a crosstabulation chart.

1. 200 people, half of whom were children, were asked if they liked ice skating. 80 replied yes, including 50% of the 60 men. Twelve of the women replied no. How many children said that they did not like ice skating?

4. 500 people, of whom 1/2 were children and 1/5 were men, were asked if they liked chewing gum. 60% said yes, including 12% of the men. 16% of the women said no. How many children liked gum?

2. 250 people, one fifth of whom were children, were asked if they used mouthwash. 160 said yes, including 90% of the 100 women. 75% of the men answered no. How many of the children used mouthwash?

5. A yes/no question was asked of 300 people. Of the 150 men, 2/3 said yes. 136 people, including 39 of the women, said no. If 25% of the yes answers were women, how many children were in the group?

3. In a group of 200 adults, 58% are women, 30% are Republicans, and 30% are Independents. Of the men, 1/3 are Republicans and 32 are Democrats. How many women are Independents?

6. Of 220 stamps in a collection, 3/11 were U.S. stamps and 50% were cancelled. If 60% of the U.S. stamps were cancelled, how many of the stamps were not U.S. and not cancelled?

Counting 4

Use a variable in the Venn to answer.

1. Of 90 students, 40 study French, 60 study Spanish, and 5 study neither. How many study both?

2. Of 130 girls, 72 wore coats, 64 wore sweaters, and 4 wore neither. How many wore both?

3. Of 160 houses, 65 need new roofs, 108 need new landscaping, and 12 need neither. How many need both?

4. Of 135 children, 95 played tennis, 50 played soccer, and 15 played neither. How any played both?

5. Of 136 students, 42 play brass instruments, 95 play string instruments, and 7 play neither. How many play both?

6. Of 73 houses, 28 need new roofs and 36 need new shrubs. If 17 need neither, how many need only a roof?

7. Of 91 students, 54 take math, 27 take music, and 13 take neither subject. How many students take only math?

8. Of 124 girls, 78 are wore sweaters, 63 wore coats, and 25 wore neither. How many wore only a coat?

9. Of 106 dresses, 76 have buttons, 79 have zippers, and 9 have neither. How many have only buttons?

10. Of 104 students, 72 take Spanish, 55 take German, and 18 take neither language. How many students take only German?

Counting 5

Count the perfect squares between the two given numbers.	*Count the numbers with the specified digit sum.*
1. between 101 and 10,010	9. How many 2-digit numbers have a perfect square digit sum?
2. between 890 and 2520	10. How many 3-digit numbers have an odd 2-digit perfect square digit sum?
3. between 601 and 3601	11. How many 2-digit numbers have a perfect cube digit sum?
4. between 50 and 8150	12. How many 3-digit numbers have an odd perfect cube digit sum?
5. between 401 and 40,001	13. How many 2-digit numbers have a 1-digit prime digit sum?
6. between 899 to 250,001	14. How many 2-digit numbers have a 2-digit prime digit sum?
7. between 48 and 490,048	15. How many 4-digit numbers have an even prime digit sum?
8. between 220 and 920	16. How many whole numbers from 100 to 400 have a digit sum of 17?

Counting 6

How many triangles?

How many squares?
How many rectangles (includes squares)?

1.

2.

3.

4.

5.

6.

7.

8.

9.

10.

11.

12.

Counting 7

Count using the Multiplication Principle. | *Count the multiples.*

1. How many 3-digit numbers with no 0s and at least one 4 are divisible by 5 ?

2. How many 3-digit numbers with different digits are divisible by 10?

3. How many 3-digit numbers have the tens digit one greater than the ones digit?

4. How many 3-digit numbers have the hundreds digit one greater than the tens digit?

5. How many 3-digit numbers with no 0s and different digits are divisible by 2?

6. How many 3-digit even numbers have all prime digits?

7. How many 3-digit even numbers have exactly 3 different prime digits?

8. How many numbers from 1 to 100 inclusive are multiples of 2 or 3?

9. How many numbers from 1 to 100 inclusive are multiples of 3 or 4?

10. How many numbers from 1 to 100 inclusive are multiples of 3 or 5?

11. How many numbers from 1 to 200 inclusive are multiples of 6 or 8?

12. How many numbers from 1 to 100 inclusive are multiples of 2, 3, or 5?

13. How many numbers from 1 to 100 inclusive are multiples of 2, 3, or 7?

14. How many numbers from 1 to 200 inclusive are multiples of 3, 5, or 7?

Counting 8

Count the number of paths from A to B moving only right and/or up.

1.

6.

2.

7.

3.

8.

4.

9.

5.

10.

Cubes 1

How many cubes with the given edge can maximally fill a rectangular box with the specified unit measurements (or be cut from a block with those measurements)?

1. cube edge 2 box (block) 8 by 10 by 12	9. cube edge 11 box (block) 55 by 22 by 66
2. cube edge 4 box (block) 36 by 24 by 20	10. cube edge 8 box (block) 32 by 16 by 64
3. cube edge 3 box (block) 36 by 18 by 15	11. cube edge 5 box (block) 15 by 35 by 50
4. cube edge 5 box (block) 45 by 30 by 25	12. cube edge 6 box (block) 24 by 30 by 66
5. cube edge 12 box (block) 24 by 48 by 60	13. cube edge 9 box (block) 24 by 48 by 60
6. cube edge 6 box (block) 36 by 18 by 60	14. cube edge 13 box (block) 26 by 52 by 65
7. cube edge 3 box (block) 6 by 6 by 9	15. cube edge 8 box (block) 24 by 40 by 72
8. cube edge 7 box (block) 14 by 42 by 56	16. cube edge 15 box (block) 45 by 75 by 90

Cubes 2

Answer as indicated. Measurements are in units, square units, and cubic units.

1. Find the volume of a cube with surface area 384.

2. Find the surface area of a cube with volume 1000.

3. Find the volume of a cube with surface area 216.

4. Find the volume of a cube with surface area 18.

5. Find the sum of the edges of a cube with volume 729.

6. Find the volume of a cube if the sum of its edges is 36.

7. Find the surface area of a cube with volume 343.

8. A cube has surface area 600. Find the sum of its edges.

9. Find the volume of a cube with surface area 150.

10. Find the surface area of a cube with volume 1728.

11. If the volume and surface area of a cube are equal, find its edge.

12. Find the volume of a cube with surface area 726.

13. Find the volume of a cube with surface area 30.

14. Find the sum of the edges of a cube with surface area 294.

15. Find the sum of the edges of a cube with surface area 96.

16. Find the surface area of a cube if the sum of its edges is 156.

Cubes 3

Answer as indicated.

1.	Find the percent change in the surface area of a cube if the edge is increased by 50%.	7.	Find the percent change in the diagonal of a cube if the volume is increased by 700%.
2.	Find the percent change in the surface area of a cube if the edge is decreased by 50%.	8.	Find the percent increase in each edge of a cube if the surface area increased by 21%.
3.	Find the percent change in the volume of a cube if the edge is increased by 50%.	9.	Find the percent change in the surface area of a cube if the edge is increased by 20%.
4.	Find the percent change in the volume of a cube if the edge is decreased by 50%.	10.	Find the percent change in the volume of a cube if the edge is increased by 20%.
5.	Find the ratio of the volumes of a cube to that of the cube with edge increased by 25%.	11.	Find the percent change in the surface area of a cube if the edge is decreased by 20%.
6.	Find the ratio of the surface areas of a cube to that of the cube with edge decreased by 33 1/3%.	12.	Find the percent change in the volume of a cube if the edge is decreased by 20%.

Cubes 4

A solid "just fits" in a cube. They have congruent heights. Find the volume inside the cube but outside the solid. Also find the ratio of the volumes, solid to cube.

1. cylinder inside cube with edge 10	7. sphere inside cube with edge 5
2. square pyramid inside cube with volume 216	8. cone inside cube with face area 4
3. sphere inside cube with face area 324	9. cylinder inside cube with edge 7
4. cone inside cube with surface area 384	10. square pyramid inside cube with face diagonal $3\sqrt{2}$
5. square pyramid inside cube with volume 729	11. cylinder with radius 3 inside cube
6. sphere inside cube with face area 16	12. cone inside cube with face area 144

Cubes 5

Find the volume in cubic units and total surface area in square units of glued solids comprised of unit cubes. Front faces are shaded only to help visualization.

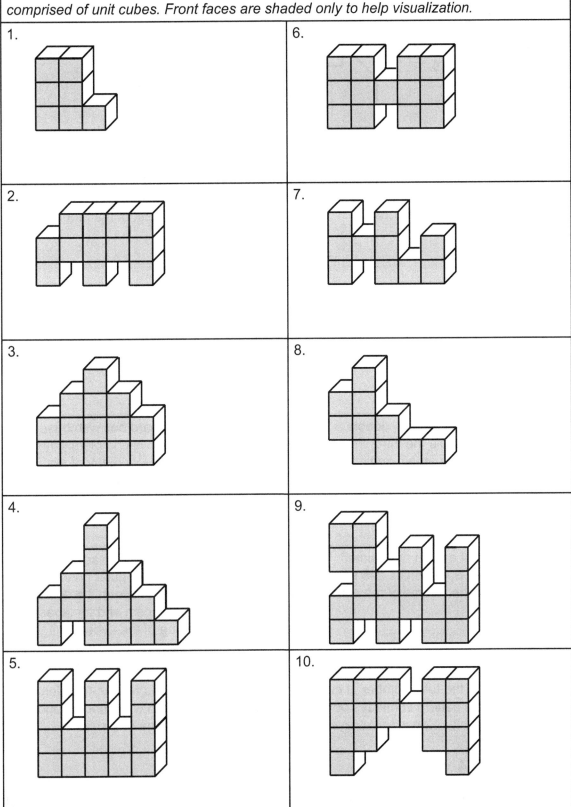

1.

2.

3.

4.

5.

6.

7.

8.

9.

10.

Cubes 6

Answer as indicated.

1. Find the surface area: cube with edge 2 on cube with edge 4.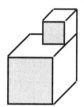

2. A cube with edge 6 is cut, perfectly centered, from the top into a cube with edge 12. Find the total surface area of the resulting figure.

3. Fix one edge of a cube, double a second, and triple the third. Find the ratios of volumes: original cube to new prism.

4. Fix one edge of a cube, double a second, and triple the third. Find the ratios of surface areas: original cube to new prism.

5. Five faces of a cube are painted. The painted area equals 245 square units. Find the volume of the cube.

6. Find the total number of cubes if the pattern continues downward for 12 horizontal layers.

7. A cube with face area 12 "just fits" inside a sphere. Find the volume of the gap between the 2 solids.

8. A cubical box with edge 1 yard is to be covered with fabric that costs $10 per square foot. Assuming no overlap and waste, find the cost of the fabric.

9. A cubical box with edge 4 is filled completely with unit cubes and then closed. How many of the unit cubes touch the cardboard?

10. A topless cardboard box with base 4 by 5 and height 6 is completely filled with unit cubes. How many of the cubes touch the cardboard?

11. When 2 unit cubes are placed face-to-face to form a rectangular prism, what is the difference between the space diagonal of the prism and the space diagonal of one of the cubes?

12. Two cubes placed face-to-face form a rectangular prism with space diagonal $2\sqrt{6}$. Find the edge of a cube.

Cubes 7

Answer as indicated for the painted cubes.

1. A cube with edge 8 is totally painted and then cut into cubes with edge 2. How many of the smaller cubes have paint on exactly 3 faces?

5. A cube with edge 4 made of 64 unit cubes is totally painted and then disassembled. For a unit cube selected randomly, find the probability it will have each number of faces painted.

    0 faces

    1 face

    2 faces

    3 faces

2. A cube with edge 5 is totally painted and then cut into 125 unit cubes. How many of the small cubes have:

0 faces painted?

1 face painted?

2 faces painted?

3 faces painted?

6. A large cube, formed by gluing unit cubes, is dipped in paint. Then it is separated into the original unit cubes, 216 of which have no painted faces. How many unit cubes were used to form the large cube?

3. A large cube, formed by gluing unit cubes, is dipped in paint and then separated into the original unit cubes, of which 294 have paint on exactly one face. Find the volume of the large cube.

7. A 4-inch cube is painted and then cut into one-inch cubes. Find the probability that a cube selected and placed randomly will have the top face painted.

4. A cube with edge n is totally painted and then cut into n^3 unit cubes. How many of the small cubes have:
0 faces painted?

0 faces painted?

1 face painted?

2 faces painted?

3 faces painted?

8. A cube with edge 3 made of 27 unit cubes is totally painted and then disassembled. For a unit cube selected randomly, find the probability it will have each number of faces painted.

    0 faces

    1 face

    2 faces

    3 faces

Cubes 8

Complete the chart.

	EDGE	VOLUME	AREA FACE	FACE DIAG	SPACE DIAG	SURFACE AREA	SUM EDGES
1.		1728					
2.						294	
3.					3		
4.						180	
5.					$\sqrt{6}$		
6.				6			
7.			15				
8.				4			
9.					$\sqrt{3}$		
10.	$\sqrt{6}$						
11.						1536	
12.						72	
13.		729					
14.					15		
15.			27				
16.				$2\sqrt{5}$			
17.			42				
18.			66				
19.				$2\sqrt{7}$			

Cylinders 1

Find the volume in cubic units of a cylinder with the given unit dimensions. Answer in terms of π.

1. height 12, radius 6	10. height 5, radius 2
2. height 30, diameter 18	11. diameter 8, height 20
3. diameter 10, height triple the radius	12. radius 3, height triple the diameter
4. radius 12, height 11	13. radius 13, height 3
5. radius 6, height half the radius	14. radius 11, height double the diameter
6. radius 8, height 22	15. diameter 28, height 10.5
7. height = 40, diameter = 18	16. diameter 8, height 11.5
8. height = 60, diameter = 22	17. radius 16, height 7
9. diameter 14, height 13	18. height = 3, diameter = 5

Cylinders 2

Find the surface area in square units of the cylinder with the given unit dimensions. Answer in terms of π.

1. height 12, radius 6	9. height 5, radius 2
2. height 20, diameter 20	10. diameter 8, height 20
3. diameter 22, height equals the radius	11. radius 3, height is twice the diameter
4. radius 9, height 11	12. radius 3, height 3
5. radius 8, height equals half the radius	13. radius 15, height is 4/5 the diameter
6. radius 10, height 40	14. radius 1.2, height 10
7. radius 5, height 7	15. diameter 5, height 9
8. height 30, diameter 3	16. height 16, radius 20

Cylinders 3

Find the volume in cubic units of the cylinder formed by rotating the rectangle with the given vertices around the specified line.

1. vertices: (0,0), (0,4), (5,4), (5,0)
   line: y-axis

2. vertices: (0,0), (0,8), (6,8), (6,0)
   line: x-axis

3. vertices: (3,0), (3,9), (7,9), (7,0)
   line: x = 3

4. vertices: (0,2), (0,6), (5,6), (5,2)
   line: y = 2

5. vertices: (−3,−2), (−3,4), (4,4), (4,−2)
   line: x = −3

6. vertices: (1,0), (1,6), (6,6), (6,0)
   line: x-axis

7. vertices: (2,−3), (2,5), (−1,5), (−1,−3)
   line: x = 2

8. vertices: (0,0), (0,7), (8,7), (8,0)
   line: x = 4

9. vertices: (1,4), (1,−6), (3,−6), (3,4)
   line: y = −1

10. vertices: (0,0), (0,6), (3,6), (3,0)
    line: y-axis

Cylinders 4

Answer as indicated. Linear measurements are in units. Volume answers are in cubic units; surface area answers are in square units.

1. Water from a full cylindrical glass with diameter 8 and height 6 is poured into a rectangular pan 4 by 6 by 9 until the pan is full. How much water remains in the glass?

2. A treasure chest is in the shape of a half cylinder on top of a rectangular box. Find the total volume if the box measures 8 deep, 10 wide, and 11 tall.

3. Find the volume between 2 cylinders, one inside the other, with concentric circular bases of diameters 9 and 12, both 10 high.

4. A cylinder has diameter 16 and height 40. A square prism with equal height to the cylinder also has equal volume. Find the side of the base of the prism.

5. A wedge is cut at a 60° central angle from a cylindrical block of cheese. Find the volume of the wedge if the cylinder has radius 8 and height 6.

6. Find the remaining volume of a cube with edge 8 after a cylindrical hole with diameter 2 is drilled through the cube.

7. Find the total surface area (inside and outside) after a cylindrical hole with diameter 4 is drilled through a cube with edge 10.

8. Find the volume of a cylindrical shell with height 40, outside diameter 10, and inside diameter 8.

Decimals 1

Solve by clearing the decimal points.

1. .9x = 81	12. 700 = 3.5x	23. 2850 = 9.5x
2. .12x = 3	13. .4x = 500	24. .17x = 51
3. 7.5 = 50x	14. 3.4x = 20.4	25. 7.25x = 58
4. 12 = .3x	15. 2.6 = .52x	26. 26.5 = 53x
5. .15x = 9	16. .06x = 19.2	27. 8.25 = 25x
6. .6x = 14.4	17. 1.2x = 60	28. 9.2 = 20x
7. .07x = 5.6	18. .008x = 0.5	29. 41.5 = 5x
8. .2x = 13	19. 18.5 = 50x	30. 242 = 5.5x
9. .025x = 3	20. 3.25 = 25x	31. 16.5 = 66x
10. 4800 = 2.4x	21. 11.2 = 20x	32. 3.25x = 39
11. .18x = 5.4	22. 23.5 = 5x	33. .11x = 264

Decimals 2

Write the exact decimal conversion.

1. $\dfrac{7}{9}$	17. $\dfrac{5}{9}$	33. $\dfrac{5}{11}$
2. $\dfrac{7}{99}$	18. $\dfrac{9}{11}$	34. $\dfrac{14}{33}$
3. $\dfrac{7}{999}$	19. $\dfrac{17}{33}$	35. $\dfrac{53}{99}$
4. $\dfrac{7}{9999}$	20. $\dfrac{4}{45}$	36. $\dfrac{1}{225}$
5. $\dfrac{7}{90}$	21. $\dfrac{23}{99}$	37. $\dfrac{1}{180}$
6. $\dfrac{7}{900}$	22. $\dfrac{35}{99}$	38. $\dfrac{2}{495}$
7. $\dfrac{7}{990}$	23. $\dfrac{47}{990}$	39. $\dfrac{31}{330}$
8. $\dfrac{5}{33}$	24. $\dfrac{83}{99}$	40. $\dfrac{1}{11}$
9. $\dfrac{1}{45}$	25. $\dfrac{1}{300}$	41. $\dfrac{25}{33}$
10. $\dfrac{3}{11}$	26. $\dfrac{19}{330}$	42. $\dfrac{32}{999}$
11. $\dfrac{13}{99}$	27. $\dfrac{6}{11}$	43. $\dfrac{25}{9999}$
12. $\dfrac{1}{90}$	28. $\dfrac{1}{55}$	44. $\dfrac{13}{33}$
13. $\dfrac{4}{9}$	29. $\dfrac{127}{9999}$	45. $\dfrac{2}{45}$
14. $\dfrac{571}{999}$	30. $\dfrac{25}{999}$	46. $\dfrac{591}{9999}$
15. $\dfrac{29}{990}$	31. $\dfrac{8}{9}$	47. $\dfrac{1}{15}$
16. $\dfrac{4}{55}$	32. $\dfrac{8}{33}$	48. $\dfrac{67}{99}$

MAVA Math: Enhanced Skills Copyright © 2015 Marla Weiss

Decimals 3

Express as a simplified fraction.	Operate on the decimals using mental math.	
1. $.2\overline{81}$	17. .2 + .93	33. 4 + .25
2. $.8\overline{24}$	18. 2 − .93	34. 4 − .25
3. $.5\overline{72}$	19. 6 + .14	35. 4 x .25
4. $.3\overline{14}$	20. 6 − .14	36. 4 ÷ .25
5. $.3\overline{27}$	21. 71 + 4.65	37. .25 ÷ 4
6. $.1\overline{25}$	22. .71 + .465	38. .8 + .064
7. $.9\overline{18}$	23. .71 + 46.5	39. .8 − .064
8. $.2\overline{63}$	24. 71 − .465	40. .8 x .064
9. $.6\overline{15}$	25. .8 + .032	41. .8 ÷ .064
10. $.5\overline{09}$	26. .8 − .032	42. .064 ÷ .8
11. $.3\overline{21}$	27. .8 x .032	43. .8 − .005
12. $.2\overline{54}$	28. .8 ÷ .032	44. .8 x .005
13. $.1\overline{36}$	29. .032 ÷ .8	45. .8 ÷ .005
14. $.3\overline{45}$	30. 12 ÷ .025	46. 20.9 − 5.06
15. $.2\overline{75}$	31. 150.8 − 14.75	47. 50.6 −2.09
16. $.3\overline{12}$	32. 1.6 ÷ .008	48. .025 x 3.2

Decimals 4

Write T if the fraction terminates as a decimal or R if it repeats. For repeating decimals, also write the denominator factor causing the repeat.

1. $\dfrac{7}{12}$	17. $\dfrac{9}{40}$	33. $\dfrac{171}{224}$
2. $\dfrac{11}{32}$	18. $\dfrac{143}{176}$	34. $\dfrac{10}{48}$
3. $\dfrac{35}{75}$	19. $\dfrac{17}{25}$	35. $\dfrac{353}{640}$
4. $\dfrac{13}{50}$	20. $\dfrac{10}{51}$	36. $\dfrac{111}{120}$
5. $\dfrac{15}{60}$	21. $\dfrac{27}{55}$	37. $\dfrac{249}{512}$
6. $\dfrac{7}{16}$	22. $\dfrac{91}{280}$	38. $\dfrac{49}{96}$
7. $\dfrac{77}{160}$	23. $\dfrac{53}{128}$	39. $\dfrac{251}{256}$
8. $\dfrac{33}{125}$	24. $\dfrac{169}{375}$	40. $\dfrac{23}{575}$
9. $\dfrac{121}{175}$	25. $\dfrac{99}{250}$	41. $\dfrac{777}{1024}$
10. $\dfrac{5}{24}$	26. $\dfrac{113}{320}$	42. $\dfrac{19}{85}$
11. $\dfrac{21}{56}$	27. $\dfrac{73}{100}$	43. $\dfrac{12}{45}$
12. $\dfrac{49}{625}$	28. $\dfrac{33}{150}$	44. $\dfrac{3421}{4000}$
13. $\dfrac{19}{80}$	29. $\dfrac{15}{64}$	45. $\dfrac{343}{750}$
14. $\dfrac{20}{57}$	30. $\dfrac{24}{25}$	46. $\dfrac{169}{800}$
15. $\dfrac{101}{680}$	31. $\dfrac{111}{440}$	47. $\dfrac{243}{700}$
16. $\dfrac{131}{275}$	32. $\dfrac{599}{600}$	48. $\dfrac{25}{52}$

Distance-Rate-Time 1

Write and solve a D=RT equation. Mph is miles per hour.

1. How many minutes does Liz spend biking 4 miles at 12 mph?

2. Find Dana's rate in mph to run 12.5 miles in 2.5 hours.

3. How many miles does Abby drive in 80 minutes at 45 mph?

4. How many minutes does Jon spend driving 12 miles at 40 mph?

5. Find Eli's rate in mph when he runs 2 miles in 12 minutes.

6. How many miles are covered at 15 mph for 24 minutes?

7. Find the rate in mph of a person who runs 4.5 miles in 45 minutes.

8. What is the rate of a car in mph that travels 1 mile in 75 seconds?

9. Kristen bikes 2 miles at 16 mph in how many minutes?

10. How many miles does a bird fly at 36 mph in 25 minutes?

11. A plane flies 300 miles at 450 mph in how many minutes?

12. What is the rate in mph of a plane that flies 90 miles in 18 minutes?

13. How many miles does a plane fly at 360 mph in 5 minutes?

14. Find a snail's rate in mph to crawl half a mile in 20 hours.

Distance-Rate-Time 2

Answer by constructing a DRT chart. Mph is miles per hour.

1. Harrison rode 60 miles at 12 mph, returning along the same route at 20 mph. What was his average rate in mph for the round trip?

2. David jogged 12 miles at 6 mph and then walked back on the same path at 2 mph. What was his average rate in mph for the round trip?

3. Amy drove 180 miles at 40 mph and returned by the same route at 60 mph. What was her average rate in mph for the whole trip?

4. Sam drove at 45 mph for 4 hours and then at 60 mph for 8 hours. What was his overall rate in mph for the entire trip?

5. A plane flew 3000 miles at 200 mph and returned by the same route at 300 mph. Find the average rate in mph for the round trip.

6. Joslyn drove 120 miles at 30 mph and returned by the same route at 60 mph. What was her average rate for the full trip?

7. Mia biked a total of 360 miles–one third at 8 mph, one half at 9 mph, and one sixth at 12 mph. What was her average rate?

8. Gil traveled 20 miles at 50 mph and the rest of his 100 mile trip at 40 mph. Find his overall rate in mph.

Distance-Rate-Time 3

Answer by constructing a DRT chart before writing and solving an equation. Mph is miles per hour.

1. At noon Mo drives north at 30 mph. Jo leaves from the same place one hour later driving north 10 mph faster to catch him. When does Jo catch Mo?

4. Kyra and Lily walked toward each other from opposite ends of a 9-mile long road. Kyra walked at 5 mph; Lily walked at 4 mph. How many miles had each walked when they met?

2. Sal and Hal left from the same place driving in opposite directions. Hal drove 12 mph faster than Sal. Two hours later they were 192 miles apart. Find Hal's rate in mph.

5. Eli paddled a canoe down a river at a steady rate of 6 mph. Two hours later Hudson left from the same place, driving a motorboat at 18 mph. After how long did Hudson reach Eli?

3. Jane walked from home to school at 3 mph and got a ride home at 42 mph. Her total travel time was 1/2 hour. How many minutes was the drive home?

6. The trip from City A to City B at 60 mph takes a half hour longer than if driving at 80 mph. How many miles is the trip?

Distance-Rate-Time 4

Answer by constructing a DRT chart, if necessary, before writing and solving a system of equations. Mph is miles per hour.

1. A boat travels upstream against the current at 7 mph and downstream with the current at 15 mph. Find the rate of the current in mph.

2. A bird flies 80 miles in 5 hours with the wind but needs 20 hours for this trip against the wind. Find the rate of the wind in mph.

3. A fish swims 240 yards downstream in 12 minutes but needs 15 minutes for the same trip in the opposite direction. Find the rate of the current in yards per minute.

4. A plane flight of 8400 miles takes 6 hours with a tail wind but 7 hours on the exact-path return flight with a head wind. Find the rate of the wind in mph.

5. A boat travels upstream against the current at 16 mph and downstream with the current at 30 mph. Find the rate of the current in mph.

6. A bird flies 1000 yards in 5 minutes with the wind, returning along the same path in 25 minutes against the wind. Find the rate of the wind in yards per minute.

7. A boater travels 90 miles downstream in 5 hours. Returning upstream, the boater's rate is 10 mph. Find the rate of the current in mph.

8. A plane flight of 360 miles takes 2 hours with a tail wind but 5 hours on the exact-path return flight with a head wind. Find the rate of the wind in mph.

Divisibility 1

Check the box if the row number is divisible by the column factor. Use divisibility rules for all.

		2	3	4	5	6	8	9	10	11	12	15	18	20	22
1.	210														
2.	264														
3.	384														
4.	396														
5.	500														
6.	528														
7.	792														
8.	900														
9.	968														
10.	1008														
11.	2640														
12.	3324														
13.	4345														
14.	4923														
15.	5082														
16.	5192														
17.	6336														
18.	6435														
19.	7260														
20.	7920														
21.	8140														

Divisibility 2

Answer as indicated using divisibility rules.

1. The four-digit number 5xx8 is divisible by 9. Find x.

2. Find all digits d so that the five-digit number 4137d is divisible by 6.

3. Find the digit D such that the five-digit number D2D3D is divisible by 11.

4. Find all values of A for which the four-digit number A55A is divisible both by 3 and 11.

5. Find the digit d so that the six-digit number 83d152 is divisible by 9.

6. Find the least digit x so that 5xx8 is a four-digit number divisible by 6.

7. Find the digit x so that 273x is a four-digit number divisible by 12.

8. How many numbers of the form 1DD2 are divisible by 6?

9. Find the digit x so that the four-digit number x2x3 is divisible by 11.

10. Find the digit x so that the four-digit number x5x8 is divisible by 11.

11. The five-digit number 12A3B is divisible by 36. Find A and B if A ≠ B.

12. Find the digit A so that the five-digit number 2A365 is divisible by 9.

13. Find the digit d so that the five-digit number 7142d is divisible by 6.

14. Find the digit x so that the six-digit number 3x2x51 is divisible by 9.

15. Find the digit x so that the 4-digit number 3xx8 is divisible by 9.

16. How many numbers of the form 1DD5 are divisible by 15?

Divisibility 3

State the remainder without dividing.

State the divisibility rule for the number.

1. $712,359 \div 5$	17. $572,815 \div 9$	33. 12 if divisible by
2. $163,967 \div 4$	18. $335,122 \div 6$	34. 14 if divisible by
3. $362,461 \div 6$	19. $350,636 \div 14$	35. 15 if divisible by
4. $427,892 \div 3$	20. $444,558 \div 9$	36. 18 if divisible by
5. $874,598 \div 10$	21. $214,845 \div 12$	37. 20 if divisible by
6. $836,733 \div 9$	22. $973,217 \div 10$	38. 24 if divisible by
7. $146,371 \div 7$	23. $376,326 \div 8$	39. 25 if ends in
8. $534,673 \div 5$	24. $174,985 \div 15$	40. 28 if divisible by
9. $697,938 \div 11$	25. $964,574 \div 6$	41. 30 if divisible by
10. $643,987 \div 2$	26. $371,245 \div 3$	42. 33 if divisible by
11. $854,593 \div 4$	27. $246,754 \div 11$	43. 35 if divisible by
12. $924,729 \div 8$	28. $745,569 \div 8$	44. 36 if divisible by
13. $208,579 \div 15$	29. $316,796 \div 5$	45. 40 if divisible by
14. $192,812 \div 11$	30. $210,753 \div 7$	46. 42 if divisible by
15. $456,147 \div 12$	31. $497,855 \div 2$	47. 44 if divisible by
16. $667,154 \div 4$	32. $495,165 \div 11$	48. 45 if divisible by

Divisibility 4

Answer as indicated for natural numbers.

1.	When a certain number is divided by 11, the quotient is 7 and the remainder is 9. Find the number.	9.	What is the greatest number by which every 6-digit palindrome must be divisible?
2.	What is the greatest number by which every 4-digit palindrome must be divisible?	10.	$6^5 \times 10^7$ can be divided by 20 how many times with no remainder?
3.	What is the greatest 2-digit number by which n must be divisible if n is divisible by 2, 3, 6, and 11?	11.	How many positive integers less than 200 are divisible by 2, 3, and 7?
4.	Find n if n is greater than 1 and less than 9, and if the remainders are the same when 9 and 16 are each divided by n.	12.	Find the least n such that n divided by 2, 3, 4, 5, and 6 has a remainder one less than the divisor.
5.	Find the product of the thousands digit and ones digit of the greatest four-digit number divisible by 4.	13.	Find the probability that a positive integer less than 1000 selected randomly is divisible by 9.
6.	Find the least n divisible by 3, 6, 7, 8, and 9.	14.	$6^6 \times 10^8$ can be divided by 30 how many times with no remainder?
7.	Find the probability that a positive integer less than 500 selected randomly is divisible by 7.	15.	Find the least n divisible by each of the first eight natural numbers.
8.	How many positive integers less than 100 are divisible by 2, 3, and 5?	16.	When a certain number is divided by 9, the quotient is 12 and the remainder is 7. Find the number.

Divisibility 5

Determine divisibility by 7 and 19 using the two rules.

1. 36,036	4. 32,604	7. 88,452	10. 35,112
2. 29,925	5. 37,908	8. 41,895	11. 26,125
3. 55,575	6. 75,803	9. 63,175	12. 87,516

Divisibility 6

Answer as indicated.

1. How many permutations of the digits 2, 3, 5, 6 result in a 4-digit number divisible by 11?	8. How many permutations of the digits 1, 3, 5, 6, 9 result in a 5-digit number divisible by 6?
2. How many permutations of the digits 0, 2, 3, 4, 9 result in a 5-digit number divisible by 90?	9. How many permutations of the digits 1, 2, 5, 7, 9 result in a 5-digit number divisible by 15?
3. How many permutations of the digits 1, 3, 5, 6 result in a 4-digit number divisible by 4?	10. How many permutations of the digits 1, 2, 4, 6, 8 result in a 5-digit number divisible by 12?
4. How many permutations of the digits 0, 4, 6, 8, 9 result in a 5-digit number divisible by 8?	11. How many permutations of the digits 0, 2, 5, 8, 9 result in a 5-digit number divisible by 18?
5. How many permutations of the digits 1, 2, 5, 6, 7 result in a 5-digit number divisible by 6?	12. How many permutations of the digits 1, 2, 3, 4, 6, 8 result in a 6-digit number divisible by 11?
6. How many permutations of the digits 1, 2, 5, 7, 9 result in a 5-digit number divisible by 25?	13. How many permutations of the digits 0, 2, 5, 7, 9 result in a 5-digit number divisible by 25?
7. How many permutations of the digits 1, 2, 3, 6, 7, 9 result in a 6-digit number divisible by 11?	14. How many permutations of the digits 2, 3, 6, 7, 8 result in a 5-digit number divisible by 8?

Divisibility 7

Find the least natural number with the given remainders.

Find the number described.

1. R 4 when divided by 5
   R 1 when divided by 3

9. between 200 and 500
   divisible by 4 and 11
   prime digit sum

2. R 1 when divided by 2
   R 2 when divided by 5
   R 5 when divided by 7

10. between 400 and 650
    divisible by 4 and 9
    digits in ascending order

3. R 3 when divided by 5
   R 3 when divided by 4
   R 1 when divided by 6

11. between 375 and 475
    divisible by 2 and 7
    digits form arithmetic sequence

4. R 2 when divided by 4
   R 2 when divided by 3
   R 1 when divided by 5

12. between 440 and 840
    divisible by 2, 5, and 9
    100s digit is 7 greater than 10s digit

5. R 7 when divided by 11
   R 3 when divided by 10
   R 1 when divided by 4

13. between 200 and 500
    divisible by 2 and 11
    palindrome
    2-digit digit sum

6. R 3 when divided by 9
   R 3 when divided by 6
   R 4 when divided by 5

14. between 200 and 400
    divisible by 12
    palindrome

7. R 5 when divided by 10
   R 1 when divided by 7
   R 3 when divided by 4

15. between 200 and 300
    divisible by 5
    digit sum 9
    not a perfect square

8. R 2 when divided by 11
   R 1 when divided by 8
   R 2 when divided by 5

16. between 400 and 600
    divisible by 5 and 7
    multiple of 10
    has odd perfect square factor, not 1

Divisibility 8

Answer as indicated. The variable x is any digit.

1. Use 2, 5, 8, 9, and x to form the greatest 5-digit number divisible by 6.

2. Use 3, 4, 7, and x to form the greatest 4-digit number divisible by 12.

3. Use 2, 4, 6, and x to form the least 4-digit number divisible by 12.

4. Use 2, 3, 6, 7, 8, and x to form the least 6-digit number divisible by 22.

5. Use 3, 5, 6, 7, 9, and x to form the greatest 6-digit number divisible by 33.

6. Use 4, 5, 8, and x to form the greatest 4-digit number divisible by 99.

7. Use 2, 4, 5, 6, and x to form the least 5-digit number divisible by 20.

8. Use 1, 2, 4, 5, 8, and x to form the greatest 6-digit number divisible by 36.

9. Find the least 3-digit number divisible by 22 subtracted from the greatest 4-digit number divisible by 14.

10. Find the least 5-digit number divisible by 35 plus the greatest 3-digit number divisible by 45.

11. Find the least 3-digit number divisible by 11 subtracted from the greatest 4-digit number divisible by 17.

12. Find the greatest 5-digit number divisible by 15 minus the greatest 3-digit number divisible by 36.

13. Find the greatest 3-digit number divisible by 3, 4, and 5 plus the least 3-digit number divisible by 3, 4, and 5.

14. Find the least 4-digit number divisible by 2, 3, and 5 plus the greatest 3-digit number divisible by 2, 5, and 7.

15. Find the least 4-digit number divisible by 3 and 10 plus the greatest 4-digit number divisible by 5 and 8.

16. Find the greatest 4-digit number divisible by 5, 6, and 11 plus the least 4-digit number divisible by 5, 6, and 11.

Equations 1

Solve mentally.

1. $x - 9 = -4$	17. $4 = 6 + x$	33. $-20x = -100$	49. $x - 12 = -5$
2. $3x = -21$	18. $x - 5 = -5$	34. $x + 13 = 2$	50. $-7x = -84$
3. $x - 2 = 1$	19. $-7x = 42$	35. $4x = -16$	51. $x - 8 = 7$
4. $-5 = 5x$	20. $350 = x + 372$	36. $5x = 400$	52. $-20 = 4x$
5. $x + 8 = 12$	21. $x + 7 = -3$	37. $0 = x + 21$	53. $x + 9 = 23$
6. $-18 = x - 11$	22. $-9 = 3x$	38. $-63 = -7x$	54. $-19 = x - 13$
7. $-8x = -40$	23. $-9 = x + 3$	39. $-19 = x + 7$	55. $-9x = -63$
8. $4 = x + 7$	24. $-9 = x - 3$	40. $12 = x - 15$	56. $22 = x + 14$
9. $-7x = 56$	25. $-27 = -3x$	41. $10x = 560$	57. $-8x = 64$
10. $x + 9 = -4$	26. $x - 42 = 42$	42. $80 = x - 15$	58. $x + 11 = -3$
11. $-2x = 0$	27. $3x = 333$	43. $11x = -99$	59. $-2x = 22$
12. $56 = -8x$	28. $x + 120 = 144$	44. $x - 15 = 32$	60. $0 = -16x$
13. $0 = x + 17$	29. $-9x = -81$	45. $4x = 484$	61. $0 = x + 16$
14. $-13x = 0$	30. $3x = 45$	46. $11x = -66$	62. $-4x = 48$
15. $72 = -6x$	31. $x + 12 = 25$	47. $x - 12 = 21$	63. $75 = -15x$
16. $0 = x + 23$	32. $-9x = -72$	48. $5x = 60$	64. $70 = x + 35$

Equations 2

Solve.

1. $\dfrac{3}{4}\,x = 36$	9. $\dfrac{2}{7}\,x = 40$	17. $\dfrac{-7}{16}\,x = 35$
2. $\dfrac{2}{5}\,x = 30$	10. $\dfrac{-4}{9}\,x = 68$	18. $\dfrac{4}{5}\,x = 60$
3. $\dfrac{4}{9}\,x = 400$	11. $\dfrac{3}{7}\,x = -24$	19. $\dfrac{-9}{20}\,x = -63$
4. $\dfrac{5}{7}\,x = -45$	12. $\dfrac{11}{15}\,x = 44$	20. $\dfrac{7}{6}\,x = 49$
5. $\dfrac{3}{8}\,x = 24$	13. $\dfrac{-5}{80}\,x = -25$	21. $\dfrac{-8}{13}\,x = 32$
6. $\dfrac{2}{3}\,x = -14$	14. $\dfrac{9}{5}\,x = -81$	22. $\dfrac{11}{12}\,x = -55$
7. $\dfrac{13}{20}\,x = 39$	15. $\dfrac{3}{5}\,x = 48$	23. $\dfrac{-7}{30}\,x = -42$
8. $\dfrac{15}{16}\,x = 75$	16. $\dfrac{12}{17}\,x = 60$	24. $\dfrac{4}{7}\,x = 28$

Equations 3

Solve. Work vertically, writing one line before the final solution.

1. $3x + 7 = 22$	12. $-6 + 7x = 57$	23. $-4x + 5 = 53$
2. $5x - 2 = 18$	13. $14 - 5x = -16$	24. $13x - 2 = 37$
3. $-4x - 5 = -29$	14. $22 + 3x = -2$	25. $-6x + 3 = 33$
4. $6 + 4x = 14$	15. $9x + 5 = 86$	26. $5 - 3x = 38$
5. $-6 - 3x = -12$	16. $-49 - 8x = 23$	27. $-3x + 2 = -19$
6. $13x - 4 = -30$	17. $13 - 2x = 7$	28. $-4x - 9 = 23$
7. $4x + 2 = 14$	18. $28 - 2x = -6$	29. $-2x + 7 = -1$
8. $15 - 3x = 30$	19. $-3 - 4x = -11$	30. $-43 - 16x = 5$
9. $7x + 2 = -47$	20. $6x - 1 = 35$	31. $-10x - 7 = 73$
10. $4x - 20 = 8$	21. $5x - 7 = 28$	32. $9x + 11 = 47$
11. $7x + 6 = 13$	22. $-x - 1 = 1$	33. $-7x - 4 = -18$

Equations 4

Simplify by adding or subtracting like terms.

1. $12x + 9x$	17. $7\pi + 5 + 9\pi + 12$	33. $12x - 20x + 16$
2. $2y - 5y$	18. $3c - 4c + 5c - 6c$	34. $15x - 55x$
3. $4\pi + 9\pi$	19. $2x - 5 + 10x + 15$	35. $3y - 2 - 5y + 19$
4. $5rs + 8rs$	20. $-4x + 7 - 3x - 3$	36. $4\pi + 2 + 6\pi - 2$
5. $6x + 8x - 3x$	21. $2x + 5 + x - 3x$	37. $3e - 7 + 7e - 2$
6. $7ab - 9ab$	22. $14abc - 20abc + 1$	38. $3 + 2x + 2 + 7x$
7. $7x + 3 + 2x + 4$	23. $(6a - 7b) + (3a + 3b)$	39. $13x - 30x + 30$
8. $7u - 13u$	24. $8m - 2m + 5m$	40. $9\pi + 5 + 9x + 11$
9. $9bc - 9b - 9c$	25. $10a + 5b + 2a + 5$	41. $4y + 8 + 3y - 11$
10. $9\pi - 20\pi + 20$	26. $(x + 2) + (x + 2)$	42. $6e + 4w + 2e + 4$
11. $16 + 5x - 11 - 2x$	27. $3 + 9g - 8 + 11g$	43. $12\pi + 5 + 3\pi - 9$
12. $22p - 17p$	28. $7 + 7x - 10 + 4x$	44. $-6 - 6y - 5 + 4y$
13. $4x + 3y + 2x + y$	29. $2ab + 3a + 4b$	45. $8rst - 20rst$
14. $4ab - 5a - 16ab + 8a$	30. $13 + 4x - 9 + 7x$	46. $13 - 9x - 20 - 2x$
15. $12xy - 30xy$	31. $-7c - 14 + 9c + 15$	47. $6\pi - 16 - 7\pi + 7$
16. $5 - 5a - 15 - 15a$	32. $9ae - 9a - 9a + 9ae$	48. $(5x - 11) + (5x - 11)$

Equations 5

Write the phrase as an algebraic expression.

1.	the sum of triple a number and five	14.	the number b decreased by ten
2.	the product of a number and six	15.	fifteen times half the number c
3.	eight times a number	16.	nine minus half the number e
4.	an even number decreased by four	17.	half the sum of a number and four
5.	five more than twice a number	18.	triple the quantity s plus two
6.	seven less than a number	19.	the positive difference of six and k
7.	one fifth of an odd number	20.	the cube of the number w
8.	one third the quantity x minus one	21.	one less than triple a number
9.	the number d increased by twelve	22.	six more than one third of a number
10.	six less than double a number	23.	the product of nine and a number squared
11.	the sum of half a number and ten	24.	the sum of four and a number cubed
12.	the product of three numbers	25.	the positive difference of p and two
13.	the square root of the sum of twice a number and four	26.	the square of the sum of one and a number

Equations 6

Express each sentence as an algebraic equation or inequality.

1. Eight more than a number is fifty.

2. Twice a number is five more than the number.

3. The positive difference of a number and three is less than seventeen.

4. The sum of four and triple a number is twelve.

5. Thirty is four more than half a number.

6. Half a number is less than or equal to nineteen.

7. Ten less than a number is fifteen.

8. One half the sum of a number and 1 is twenty.

9. One third a number increased by six is thirty.

10. Four is seven less than six times a number.

11. A number squared is three more than half the number.

12. One fifth a number is eighty.

13. Five times the sum of a number and three is sixty-two.

14. A number decreased by five is one fourth the number.

15. Nine times a number, then decreased by six, is sixteen.

16. Nine times the quantity of a number decreased by six is sixteen.

17. Twice a number decreased by one is greater than triple the number.

18. Five is greater than double a number.

19. Nine times a number is less than two.

20. The sum of a number squared and ten is triple the number.

21. The positive difference of triple a number and two is the number.

22. A number cubed increased by eight is the number.

Equations 7

Solve, simplifying each side before combining sides.

1. $14x + 1 - x + 6 = 5 - 6x - 6 + 3x$	8. $-4x + 9 - 3x - 13 = 2x - 4 + x - 5$
2. $-2 + 4x + 7 - 2x = -3x + 9 + 7x$	9. $3x - 3 - 8x - 9 = -2x - 15 - x + 11$
3. $15 - 10x + 6 - x = 3x - 1 - 2x + 4$	10. $8x + 19 - 11x - 18 = 7x - 7 - 3x - 1$
4. $13x - 9 + x - 3 = -8x - 9 - 2x + 6$	11. $-2x + 10 - 5x - 2 = 5x + 23 - 8x - 11$
5. $-7 + 5x - 8 + 2x = 3 - 6x - 4 + x$	12. $5 + 8x + 6 - 3x = 4x - 30 - 19x + 19$
6. $7x - 7 + 5x - 2 = -7 + 4x - 2 - 9x$	13. $30x - 20 - 38x + 9 = 15 + 5x - 8 + x$
7. $12x - 6 - 5x + 5 = 3x + 4 - 9x + 2$	14. $8x - 6 - 9x - 1 = 3x - 9 - 7x + 17$

Equations 8

Distribute.

	Distribute. Combine like terms.
1. $3(2a + 3b + 9c)$	12. $2(x + 2y) + 5(4x + 5y)$
2. $4(8x - 5y - 3)$	13. $6(4x - 2y) + 3(3x - 4y)$
3. $-5(6x - 2y + 8)$	14. $-7(-2a - 5b) - 6(4a + 7b)$
4. $-7a(4a - 8b + 10)$	15. $3(2a + 5b - c) - 2(a - 2b + 7c)$
5. $-2(-3w - 4x + 5xw - 5)$	16. $-4(3x - 2y) - 5(2x - 5y)$
6. $-6w(5w + 2x - 12)$	17. $-8(6x - 2y) - 9(3x - 6y)$
7. $3ab(4b^3 + 5ab - 2a + 7b)$	18. $14x(2x - 5) - 5x(3x + 2)$
8. $-8c(4cd - 5c + 11)$	19. $-3a(4a + 10) - a(2a + 9)$
9. $12c(4de - 7e + 11c^3)$	20. $-5e(4e + 3x) - 4e(3e - 7x)$
10. $-9b^2(2b^6 - 5b^3 - 7b^2 + 3)$	21. $15(a + 2b - 3c) - 12(2a - b + c)$
11. $11x^3(-4x^3 - 2x^2 - 5x + 8)$	22. $3(4w - 8y + 6) - 2(6w - 12y + 9)$

Equations 9

Write a variable expression.

1. the value in cents of d dimes	14. the number of days in x leap years
2. Jan's age in 5 years if Jan is y years old now	15. the next odd number after the odd number d
3. the number of eggs in x dozen	16. the number of days in h hours
4. the number of quarts in g gallons	17. the number of hours in d days
5. the number of gallons in q quarts	18. the value in cents of q quarters
6. the number of hours in y minutes	19. the supplement of a° in degrees
7. the number of minutes in y hours	20. the money earned in dollars in d days at y dollars per day
8. the value in cents of n nickels	21. twice the complement of a° in degrees
9. the number of cups in q quarts	22. the next even number after the odd number d
10. the number of quarts in c cups	23. the value in cents of h half dollars and p pennies
11. the number of hours in w weeks	24. the next even number after the even number e
12. the number of cents in d dollars	25. the next natural number after the natural number n
13. the sum of two consecutive whole numbers	26. the money earned in dollars in h hours at x dollars per hour

Equations 10

Solve.

1. $-5(x - 6) - 6 = 10 - 2(x - 1)$	7. $7 + 6x - 5(x - 2) = -3x - (-5 - x)$
2. $9 - 4(2x - 3) = 6(x - 2) + 5$	8. $-2(5x - 3) - 2x + 12 = 9 + 6(3x - 1)$
3. $3x + 1 - 2(x - 6) = -3(x - 1) + 2$	9. $-3(x - 5) - 4x + 2 = -5(x - 2) - 3$
4. $3 - 7(2x - 4) + 3x = -6(x - 1) + 5$	10. $(-3x)(-5) + 4(x - 6) = -(x - 5) + 9 + x$
5. $-6(3x - 2) - 7 - x = 10 - 3(x - 5) + 4x$	11. $-4(2x - 4) - 2x + 9 = 5x - 9 - (x + 8)$
6. $7x + 5(x - 3) + 5 = -3(3x - 3) + 2x$	12. $-2(4x - 7) + 6x + 4 = 17 + 8(2x - 1)$

Equations 11

Solve.

1. $\dfrac{3}{4}x - 6 = 7x + \dfrac{2}{3}$

2. $\dfrac{3}{4} - 5x = 10 + \dfrac{7x}{6}$

3. $\dfrac{5}{8} - 2x = 6 + \dfrac{3x}{20}$

4. $\dfrac{5}{6}x + 5 + \dfrac{2x}{3} = 3x + \dfrac{1}{2} - \dfrac{4x}{5}$

5. $\dfrac{2}{15}x - 4 + \dfrac{x}{6} = 5x + \dfrac{1}{2} - \dfrac{2x}{5}$

6. $\dfrac{9}{10}x + 3x + \dfrac{5}{2} = 13 + \dfrac{3}{8}x$

7. $\dfrac{11}{3}x - 3 = 4 + \dfrac{2}{5}x$

8. $\dfrac{5}{8x} + 2 = 9 - \dfrac{5}{6x}$

9. $\dfrac{7}{3}x - 5 = 9x + \dfrac{5}{6}$

10. $\dfrac{5}{9x} + 2 = \dfrac{10}{3x} - \dfrac{1}{x}$

11. $\dfrac{3}{4}x - 2 = 3x + \dfrac{5}{14}$

12. $\dfrac{5}{6x} + 3 = 8 - \dfrac{5}{9x}$

Equations 12

Solve the system by substitution.

1. $X = 9$ $X - Y = 20$	8. $X + Y = 5$ $2X - 3Y = 25$	15. $3M + 4N = 4$ $2M + N = 6$
2. $P = 15$ $5Q + 2P = 60$	9. $P + 3Q = -7$ $3P - 2Q = 1$	16. $4X + 3Y = 11$ $4X - Y = 7$
3. $9X = 63$ $3X - 2Y = 71$	10. $X + Y = 9$ $2X + 5Y = 9$	17. $4P - Q = 9$ $P - 3Q = 16$
4. $4Y + 1 = 17$ $-11X + 5Y = 75$	11. $X + 2Y = 6$ $3X - 4Y = 8$	18. $X + 3Y = 1$ $-5X + Y = 11$
5. $M = N$ $4M + 7N = 88$	12. $X - Y = -1$ $2X - 3Y = 0$	19. $3X - Y = 1$ $9X - 2Y = 5$
6. $15 - 3Y = 6$ $X - 4Y = 38$	13. $4M - N = 2$ $10M - 3N = 4$	20. $3X + Y = 7$ $7X - 3Y = 43$
7. $-6A + 5 = 17$ $5A + 3B = 41$	14. $A + B = 7$ $2A - B = 8$	21. $X + 4Y = 10$ $5X + 6Y = 8$

Equations 13

Solve the system by addition/subtraction.

1. $M + N = 28$
$M - N = 14$

2. $7X - 3Y = 66$
$2X - 3Y = 21$

3. $6X - 9Y = 60$
$6X - 2Y = 18$

4. $9C + 3D = -3$
$2C - 3D = -8$

5. $X - 4Y = 13$
$X - 7Y = 70$

6. $2A - 6B = 11$
$7A + 6B = 52$

7. $3M + N = -6$
$2M - N = 1$

8. $4X - 3Y = 26$
$2X - 3Y = 16$

9. $3X - 9Y = -51$
$-3X + 5Y = 15$

10. $5C + 6D = 37$
$2C - 6D = 12$

11. $2X - 3Y = -15$
$2X - 9Y = -33$

12. $3A - B = 0$
$A + B = 4$

13. $A + B = 38$
$A - B = 20$

14. $5X - 4Y = 64$
$3X - 4Y = 32$

15. $5X - 7Y = 60$
$5X - 6Y = 45$

16. $8C + 7D = -4$
$2C - 7D = -6$

17. $M - 4N = 42$
$M - 8N = 90$

18. $A - B = 4$
$2A + B = 5$

Equations 14

Multiply by the "FOIL" method.

1. $(x + 5)(x + 4)$	12. $(2x + 9)(x + 3)$	23. $(4x + y)(x + 6y)$
2. $(x + 3)(x + 7)$	13. $(3x + 2)(4x + 5)$	24. $(2x + 4y)(3x + 5y)$
3. $(x - 6)(x - 2)$	14. $(x - 5)(4x - 7)$	25. $(8x + 3y)(7x - 2y)$
4. $(x - 7)(x - 8)$	15. $(11x - 2)(3x - 1)$	26. $(5x - 3y)(3x + 5y)$
5. $(x + 6)(x - 3)$	16. $(6x - 5)(2x + 4)$	27. $(x - 2y)(11x - 9y)$
6. $(x - 9)(x + 4)$	17. $(10x + 3)(x - 2)$	28. $(7x - 5y)(6x - y)$
7. $(x + 11)(x + 5)$	18. $(7x + 2)(3x + 2)$	29. $(4x + 3y)(2x + 8y)$
8. $(x + 7)(x - 10)$	19. $(x - 9)(8x + 4)$	30. $(11x + 5y)(2x - 6y)$
9. $(x - 4)(x + 13)$	20. $(9x - 4)(5x - 6)$	31. $(8x - 4y)(10x + 3y)$
10. $(x - 8)(x - 3)$	21. $(4x + 11)(2x - 3)$	32. $(2x - 7y)(9x - 8y)$
11. $(x + 12)(x + 5)$	22. $(2x - 7)(3x - 8)$	33. $(7x + 4y)(7x - 4y)$

Equations 15

Make a chart using one variable before writing equations and solving.

1. Sam scored twice as many points at the Sunday game as the Friday game. Rob scored the same as Sam on Friday but half as many as Sam on Sunday. If their total points were 60, find the number Sam scored on Sunday.

2. Val swam 5 times as many races as she ran. Hal ran 4 fewer races than Val swam and swam 3 more than Val ran. If they entered 35 total races, find Hal's number of running races.

3. Pam earned half as much writing as she did teaching. Bob's teaching earnings were $3000 less than Pam's, but his writing earnings were $4000 more. If their total earnings were $85,000, find Bob's teaching earnings.

4. A bus used the same amount of gas all 7 days one week. A taxi used half as much as the bus on the weekend days and one third as much as the bus on the weekdays. If 580 gallons of gas were used in total, find the taxi's weekend gas used in gallons.

5. The number of students in Lara's BA class was 150 fewer than Greg's. Her MBA class had half as many as her BA class, while Greg's MBA class had 300 fewer than his BA class. If the total number enrolled was 980, find the number in Lara's MBA class.

6. On Tuesday Bo worked triple his Friday hours, while Jo worked double Bo's Friday hours. On Friday Jo worked half of Bo's Tuesday hours. If their total hours for the 2 days is 90, find Bo's Tuesday hours.

7. One week, Meg walked 6 times as many miles as she ran. Peg ran twice as many miles as Meg ran but only half the number that Meg walked. Their total combined miles are 102. Find the number of miles that Peg ran.

8. Lil baked 4 times as many cocoa cookies as sugar cookies. Jon baked half as many sugar cookies as Lil and one third as many cocoa cookies. The total number of cookies baked was 820. Find the number of sugar cookies baked.

Equations 16

Answer as indicated by writing and solving equations.

1. There are triple as many A as C, half as many C as B, and 55 A and B combined. Find A − (B + C).

5. a + b + c + d = 500
a + b = 185
b + c = 290
a + c = 345
Find all 4 variables.

2. A total of 200 tickets were sold, some adult at $60 each and some student at $15, yielding $5700 income. How many of each type were sold?

6. Rob's items are worth $10 each, while Sue's are worth $25 each. How many does each have if together they have 150 at a total value of $2325?

3. Jo, Bo, Mo, and Ro are dividing $175,000. Mo gets half as much as Jo. Bo gets 3/4 of Jo's share. Ro gets 5,000 less than Bo. Find how much each received.

7. Hal's age is 9 more than half Tom's age. Bob's age is 5 more than one-third Tom's age. Hal is fourteen years older than Bob. Find their ages.

4. e + f + g + h = 450
e + g = 160
e + f = 90
f + g = 120
Find h.

8. The average of 3 numbers is 40. The first is twice the second. The third is 5 more than the first. Find the numbers.

Equations 17

Solve the system with multiplication followed by addition/subtraction.

1. $6A + 5B = 22$
      $12A - 7B = 10$
	7. $9M + 5N = -3$
	      $5M - 9N = 69$
		13. $4M + 5N = 11$
		       $5M - 4N = 24$

2. $4X + 3Y = 22$
      $8X - 7Y = 70$
	8. $5X + \ Y = \ 8$
	      $3X - 4Y = 14$
		14. $5X + 3Y = 19$
		       $2X - 5Y = 20$

3. $3X + 5Y = 65$
      $4X - 3Y = -10$
	9. $5X + 7Y = 6$
	      $10X - 3Y = 46$
		15. $6X + 7Y = 8$
		       $7X - 2Y = 50$

4. $6X - 7Y = 50$
      $3X + 5Y = 8$
	10. $2P + 3Q = 12$
	       $3P - 4Q = 1$
		16. $5X - 3Y = 24$
		       $3X + 5Y = 28$

5. $10B + 3C = 45$
      $5B + 3C = 15$
	11. $2X + 3Y = 20$
	       $5X - 6Y = 23$
		17. $2B + 3C = 12$
		       $5B + 7C = 29$

6. $5M - 2N = 23$
      $4M - 3N = 3$
	12. $4X + 3Y = -1$
	       $2X + 5Y = \ 3$
		18. $2M - 3N = -7$
		       $5M + 4N = 17$

Equations 18

Solve the quadratic equation in factored form. Use mental math.

1. $(x + 7)(x - 5) = 0$	17. $(2x + 3)(6x + 5) = 0$
2. $(x + 2)(x + 9) = 0$	18. $(7x + 9)(8x - 7) = 0$
3. $(x - 1)(x - 8) = 0$	19. $(7x - 4)(4x + 1) = 0$
4. $(x - 4)(x + 10) = 0$	20. $(5x - 8)(3x - 2) = 0$
5. $(x + 3)(x - 11) = 0$	21. $(4x - 3)(6x + 11) = 0$
6. $(x + 1)(x + 4) = 0$	22. $(5x - 9)(9x - 7) = 0$
7. $(x - 7)(x - 9) = 0$	23. $(3x + 4)(2x - 9) = 0$
8. $(x - 10)(x + 8) = 0$	24. $(9x + 5)(8x + 1) = 0$
9. $(2x + 1)(7x - 2) = 0$	25. $(8x - 5)(2x - 5) = 0$
10. $(9x + 4)(8x + 3) = 0$	26. $(4x + 7)(5x - 2) = 0$
11. $(4x - 5)(5x - 6) = 0$	27. $(3x - 1)(10x + 3) = 0$
12. $(3x - 8)(6x + 7) = 0$	28. $(7x + 3)(9x + 2) = 0$
13. $(8x + 9)(3x - 10) = 0$	29. $(11x + 4)(6x - 1) = 0$
14. $(6x + 1)(7x + 4) = 0$	30. $(9x - 5)(4x - 11) = 0$
15. $(2x - 7)(4x - 9) = 0$	31. $(3x - 5)(12x + 5) = 0$
16. $(9x - 10)(5x + 8) = 0$	32. $(10x - 1)(3x + 7) = 0$

Estimating 1

Estimate to the closest multiple of 5.

1.
$$\frac{(5.19)(4.92)}{0.51}$$

2.
$$\frac{(51.9)(4.92)}{0.51}$$

3.
$$\frac{(0.519)(49.2)}{5.1}$$

4.
$$\frac{(8.14)(7.88)}{0.81}$$

5.
$$\frac{(0.984)(88.8)}{9.1}$$

6.
$$\frac{(81.4)(7.98)}{0.81}$$

7.
$$\frac{(0.614)(986.721)}{(1.93)(31.1)}$$

8.
$$\frac{(0.886)(99.81)(6.52)}{(5.04)(9.07)(1.32)}$$

9.
$$\frac{(89.7)(74.65)(0.11)}{(2.98)(0.317)(24.895)}$$

10.
$$\frac{(6.03)(0.491)(51.89)}{(4.04)(0.59)(12.98)}$$

11.
$$\frac{(78.791)(71.921)(0.3)}{(5.87)(7.99)(1.188)}$$

12.
$$\frac{(6.96)(49.88)(0.89)(48.9)}{(4.984)(69.9)(8.8)}$$

13.
$$\frac{(10.89)(6.941)(14.03)(0.497)}{(7.02)(2.16)(0.91)}$$

14.
$$\frac{(0.507)(29.82)(6.61)(4.882)}{(6.06)(0.29)(10.98)(0.991)}$$

15.
$$\frac{(8.896)(4.501)(10.95)(0.391)}{(11.01)(0.891)(0.402)}$$

16.
$$\frac{(0.516)(98.721)(3.87)(10.001)(0.686)}{(4.04)(9.9)(0.498)(6.498)(1.001)}$$

Estimating 2

Answer as indicated.	Estimate to the nearest multiple of 10.
1. Eli has 159 pictures to put in an album. If each page holds 15 photos, how many pages will he need?	8. 4478 − 92
2. Weight at age one is about triple the birthweight. If a baby weighed 7 pounds 7 ounces at birth, find the weight at age one to the nearest half pound.	9. 5798 + 94
3. Would 3 school buses, each holding 40 people plus the driver, be enough to transport 6 classes of 18 students each plus 1 teacher per class?	10. 3877 + 54
4. Hudson drove 1496 miles during a five-day trip. About how many miles did he average each day?	11. 9218 − 2181
5. Harrison's annual salary is $72,250. How much does he earn each month, to the nearest ten dollars?	12. 5602 − 2818
6. A room and carpet are both 12 feet wide. The room is 11 feet long. The carpet costs $44.95 a running yard and may be cut to any length. About how much will the carpet cost?	13. 72,596 ÷ 597
7. Kyra wishes to buy a jacket that costs $59.95. If she saves $2.50 each week, in how many weeks will she have enough money for the purchase?	14. 9011 ÷ 89

Evaluating 1

Evaluate.

1. $90 - 3xy + x^2$ for $x = 3$, $y = 6$

2. $100 - 2(x + 4)^2$ for $x = 3$

3. $3(x - y) + 2xy$ for $x = 7$, $y = 3$

4. $3xy - 3x - 3y$ for $x = -2$, $y = 4$

5. $-5x - 2x^2 + 4y$ for $x = 4$, $y = -6$

6. $7x(x - y) + 3y$ for $x = 2$, $y = 5$

7. $8x - 2(2x + y)$ for $x = -2$, $y = 6$

8. $2xy + x^y + y^x$ for $x = 3$, $y = 4$

9. $x^2 + y^3 + 4xy$ for $x = 2$, $y = 3$

10. $y^2(3xy - 2y - 6)$ for $x = -2$, $y = -3$

11. $\dfrac{x - y}{x + y}$ for $x = 4$, $y = 10$

12. $3xz - 2x + 4yz$ for $x = 2$, $y = 3$, $z = 5$

13. $z^2 + 2yz - 5y$ for $y = 4$, $z = -5$

14. $5xy(x - z)$ for $x = 2$, $y = 3$, $z = -4$

15. $xy + yz - xz$ for $x = 3$, $y = 4$, $z = -2$

16. $xy - 4x + 5z$ for $x = 6$, $y = -7$, $z = 2$

17. $2y^2(5xy - 2x - 4)$ for $x = -2$, $y = -4$

18. $8x - (xy - z)$ for $x = 5$, $y = -3$, $z = 7$

19. $2z^y + 5y^2 - 2z$ for $y = 3$, $z = -5$

20. $\dfrac{4x}{4y - x}$ for $x = 6$, $y = 2$

21. $\dfrac{8xy}{5(x - y)}$ for $x = 2$, $y = 10$

22. $\dfrac{2x - 3y}{4x + 2y}$ for $x = 3$, $y = -3$

Evaluating 2

Evaluate.

1. $20 - 12xy$ for $x = \dfrac{3}{4}$, $y = \dfrac{2}{3}$	9. $xy + 4x - y$ for $x = \dfrac{3}{8}$, $y = \dfrac{2}{5}$
2. $100x - 25(x + 1)^2$ for $x = \dfrac{3}{5}$	10. $xy - 3x + 2y$ for $x = \dfrac{2}{9}$, $y = \dfrac{3}{4}$
3. $28(y - x) + 7xy$ for $x = \dfrac{1}{4}$, $y = \dfrac{4}{7}$	11. $16xy(y - x)$ for $x = \dfrac{3}{8}$, $y = \dfrac{2}{3}$
4. $27xy - 12x - 9y$ for $x = \dfrac{5}{6}$, $y = \dfrac{4}{9}$	12. $6xz - 2x + 3z$ for $x = \dfrac{1}{6}$, $z = \dfrac{11}{12}$
5. $-5x - 16y + 4xy$ for $x = \dfrac{3}{5}$, $y = \dfrac{5}{8}$	13. $3z - 5yz - 5y$ for $y = \dfrac{3}{5}$, $z = \dfrac{7}{9}$
6. $24x - 2(2x + y)$ for $x = \dfrac{5}{8}$, $y = \dfrac{3}{4}$	14. $3y^2(6xy + 2x)$ for $x = \dfrac{3}{4}$, $y = \dfrac{1}{3}$
7. $6x(3x + 2y) + 4y$ for $x = \dfrac{5}{3}$, $y = \dfrac{5}{2}$	15. $y^2(3xy - 2x)$ for $x = \dfrac{5}{6}$, $y = \dfrac{3}{2}$
8. $-18x - 7y + 21xy$ for $x = \dfrac{-2}{9}$, $y = \dfrac{3}{7}$	16. $3y^3 + 5x^2 - xy$ for $x = \dfrac{2}{5}$, $y = \dfrac{1}{3}$

Evens & Odds 1

Answer as indicated.

1. If 16 is written as the product of 3 positive integers, each greater than 1, how many of them are even?	7. Remove 5 even numbers from the first 55 whole numbers. What percent of the remaining numbers are even?
2. If the sum of two consecutive odd numbers is 28, what is their product?	8. Remove 9 odd numbers from the first 40 whole numbers. What percent of the odd numbers were removed?
3. If x is an odd number, what is the sum of the next two odd numbers greater than 3x+1?	9. If x is an even number, what is the sum of the next four odd numbers greater than 2x+3?
4. Find the least positive integer n such that 3n is both even and a perfect square.	10. If the product of the digits of a two-digit number is odd, is the sum of the digits even or odd?
5. Define n*=3n if n is even and n*=2n if n is odd. Find a* + b* if a and b are the first two prime numbers.	11. If 210 is written as the product of 3 positive integers, each greater than 1, how many of them are even?
6. Subtract the sum of the odd numbers 1 through 39 from the sum of the even numbers 2 through 40.	12. The results of 3/4 of an even number and 2/3 of the consecutive even are consecutive and descending. Find the odd number between the two evens.

Evens & Odds 2

Operate as indicated on natural numbers.	*Find the greatest number that must be a factor of the given sum.*
1. the difference of the 231st odd number and the 159th even number	9. the sum of 2 consecutive odd numbers
2. the sum of the 105th even number and the 175th odd number	10. the sum of 3 consecutive odd numbers
3. the quotient of the 210th even number and the 95th odd number	11. the sum of 4 consecutive odd numbers
4. the product of the 55th even number and the 43rd odd number	12. the sum of 5 consecutive odd numbers
5. the sum of the 195th even number and the 243rd odd number	13. the sum of 2 consecutive even numbers
6. the quotient of the 116th odd number and the 132nd even number	14. the sum of 3 consecutive even numbers
7. the product of the 100th even number and the 345th odd number	15. the sum of 4 consecutive even numbers
8. the difference of the 318th even number and the 131st odd number	16. the sum of 5 consecutive even numbers

Exponents 1

Evaluate by mental math.

1. 2^3	17. 8^2	33. 2^7	49. 700^2
2. 3^2	18. 100^2	34. 196^1	50. 90^2
3. 10^2	19. 3^4	35. 2^{10}	51. 70^2
4. 40^2	20. 2^8	36. 110^2	52. 2^5
5. 20^2	21. 5^3	37. 80^2	53. 150^2
6. 6^2	22. 30^3	38. 11^3	54. 781^0
7. 2^6	23. 60^2	39. 3^5	55. 2^{11}
8. 12^2	24. 3^6	40. 4^4	56. 30^3
9. 600^2	25. 10^4	41. 1^{82}	57. 2^2
10. 3^3	26. 0^{24}	42. 5^4	58. 11^4
11. 11^2	27. 20^3	43. 20^4	59. 90^2
12. 4^2	28. 4^5	44. 7^2	60. 50^2
13. 15^2	29. 10^3	45. 50^3	61. 56^0
14. 9^3	30. 1^{37}	46. 4^6	62. 120^2
15. 13^2	31. 14^2	47. 2^9	63. 2^4
16. 4^3	32. 300^2	48. 9^2	64. 0^{86}

Exponents 2

Evaluate.

1. $.2^2$	17. $.17^2$	33. $.2^7$
2. 1.3^2	18. $.001^2$	34. 1.1^3
3. $.02^3$	19. 1.1^2	35. $.2^{10}$
4. $.05^2$	20. $.09^3$	36. $.2^4$
5. $.02^4$	21. $.1^5$	37. $.8^2$
6. $.6^2$	22. $.5^3$	38. $.3^4$
7. $.2^6$	23. $.3^3$	39. $.14^2$
8. 1.2^2	24. $.6^3$	40. $.4^4$
9. $.5^2$	25. 1.6^2	41. $.0965^1$
10. $.3^3$	26. $.018^2$	42. $.05^4$
11. $.11^2$	27. $.12^2$	43. $.3^2$
12. $.004^2$	28. 1.7^2	44. $.7^2$
13. $.15^2$	29. $.04^3$	45. $.01^3$
14. $.01^4$	30. $.1^3$	46. $.09^2$
15. $.1^2$	31. $.011^2$	47. $.2^3$
16. $.011^2$	32. 1.4^2	48. $.4^2$

Exponents 3

Simplify using rules of exponents. Answer in exponential form.

1. $3^8 \cdot 3^6$	17. $4^8 \div 4^2$	33. $18^4 \cdot 18^9$	49. $14^7 \cdot 14^{13}$
2. $7^2 \cdot 7^5$	18. $6.2^6 \cdot 6.2^2$	34. $12^3 \cdot 12^{11}$	50. $8^9 \div 8^7$
3. $4^{20} \div 4^{10}$	19. $7^9 \div 7^5$	35. $15^{14} \div 15^7$	51. $17^{13} \div 17^4$
4. $3^6 \div 3^3$	20. $67^7 \cdot 67^8$	36. $\left(9^9\right)^2$	52. $\left(39^7\right)^4$
5. $\left(3^4\right)^2$	21. $\left(11^9\right)^3$	37. $\left(6.1^5\right)^7$	53. $11^{11} \div 11^8$
6. $\left(5^5\right)^3$	22. $22^{14} \cdot 22^3$	38. $8^{15} \div 8^9$	54. $33^9 \cdot 33^9$
7. $2^5 \cdot 2^6$	23. $25^{12} \div 25^8$	39. $\left(2^7\right)^6$	55. $21^{16} \div 21^2$
8. $9^9 \div 9^5$	24. $\left(32^5\right)^6$	40. $43^9 \cdot 43^{12}$	56. $\left(88^9\right)^7$
9. $\left(2^3\right)^3$	25. $12^3 \cdot 12^3$	41. $6^{11} \div 6^6$	57. $71^6 \cdot 71^7$
10. $17^4 \cdot 17^8$	26. $5^2 \cdot 5^6$	42. $.7^4 \cdot .7^{10}$	58. $8^{21} \div 8^{17}$
11. $7^8 \div 7^3$	27. $26^{13} \div 26^7$	43. $\left(16^8\right)^2$	59. $\left(64^5\right)^{12}$
12. $16^3 \cdot 16^4$	28. $\left(10^8\right)^4$	44. $37^{29} \div 37^{15}$	60. $92^9 \cdot 92^{21}$
13. $\left(7^2\right)^5$	29. $14^5 \cdot 14^6$	45. $\left(2.9^4\right)^4$	61. $31^{15} \div 31^8$
14. $25^6 \div 25^4$	30. $\left(12^3\right)^{10}$	46. $\left(53^4\right)^6$	62. $16^9 \div 16^7$
15. $\left(1.5^5\right)^4$	31. $\left(16^7\right)^2$	47. $1.5^9 \cdot 1.5^7$	63. $\left(35^9\right)^9$
16. $1.9^3 \cdot 1.9^6$	32. $19^{15} \div 19^6$	48. $9^{10} \div 9^4$	64. $\left(77^2\right)^{18}$

Exponents 4

Simplify using rules of exponents. Assume no denominator variable is zero.

1. $x^2 \cdot x^7$	17. $wb^6 \cdot w^5 b^7$	33. $25b^9 \div (5b^2)$
2. $y^{18} \div y^{12}$	18. $h^5 \cdot h^6 \cdot h^4$	34. $25b^9 \div (5b)^2$
3. $h^8 \div h^3$	19. $(g^2)^9$	35. $36c^{11} \div (3c)^2$
4. $(g^4)^3$	20. $eg^{10} \cdot ge^{10}$	36. $36c^{11} \div 3c^2$
5. $(ac^5)^7$	21. $ab^{12} \div b^9$	37. $(7w^9)^2$
6. $p^6 \cdot p^5$	22. $(a^4 b)^4$	38. $(3h^6)^3$
7. $x^9 \div x^8$	23. $n^2 \cdot n^7 \cdot n^5$	39. $a^9 \cdot a^3 \cdot a^8$
8. $(xy^3)^3$	24. $g^{14} \cdot g^6$	40. $(3c^7)^4$
9. $m^4 \cdot m^9$	25. $x^2 y^{13} \div y^8$	41. $x^5 \cdot ax^5 \cdot bx^2$
10. $s^{11} \div s^3$	26. $(w^4)^6$	42. $(6e^5)^2$
11. $a^4 b^3 \cdot a^5 b^4$	27. $eh^2 \cdot he^2 \cdot h^6$	43. $(2g^4)^4$
12. $(y^3)^5$	28. $(x^3)^9$	44. $(k^2 s^4)^3$
13. $e^9 \div e^2$	29. $(p^7)^3$	45. $c \cdot cn^6 \cdot nc^3$
14. $(c^2 d^5)^4$	30. $g^{15} \div g^{14}$	46. $m^4 \cdot a^4 \cdot m^4$
15. $c^2 d^5 \cdot a^2 d^6$	31. $a^3 \cdot a^8 \cdot a^6$	47. $y^{30} \div y^{20}$
16. $k^{13} \div k^5$	32. $a^2 \cdot an^7 \cdot na^5$	48. $e^7 c^{15} \div e^4$

Exponents 5

Write as the least whole number times a power of ten.	Operate and simplify. Answer as powers of x and 2, 5, and/or 10.
1. $(4 \cdot 10^2) \cdot (5 \cdot 10^2)$	14. $(5x^2)^4(2x^3)^4(2x^6)(5x^7)$
2. $(200 \cdot 10^5) \div (5 \cdot 10^3)$	15. $(5x)(4x^2)^2(2x^5)(4x^2)^3(2x^6)$
3. $(15 \cdot 10^2) \cdot (4 \cdot 10^6)$	16. $(2x^3)^2(8x^5)(2x^2)^3(4x^5)$
4. $(5 \cdot 10^5) \cdot (16 \cdot 10^4)$	17. $(2x^5)^3(5x^5)(2x^2)^3(5x)^5$
5. $(80 \cdot 10^{11}) \div (8 \cdot 10^4)$	18. $(10x^2)(5x^5)(2x^2)^4(25x^5)^2$
6. $(120 \cdot 10^8) \div (4 \cdot 10^3)$	19. $(2x^3)^3(8x^6)(2x)^3(5x^5)$
7. $(18 \cdot 10^5) \cdot (5 \cdot 10^5)$	20. $(4x^2)^2(2x^5)(5x^2)^3(5x^5)^2$
8. $(80 \cdot 10^{12}) \div (2 \cdot 10^4)$	21. $(5x^2)^3(125x^5)(2x^2)^4(40x^5)$
9. $(14 \cdot 10^6) \cdot (5 \cdot 10^7)$	22. $(4x^4)^2(16x^3)(4x^3)^2(8x^7)$
10. $(500 \cdot 10^7) \div (25 \cdot 10^6)$	23. $(25x^3)(8x^6)(10x^3)^3(50x^4)$
11. $(15 \cdot 10^3) \cdot (8 \cdot 10^{11})$	24. $(2x^4)^4(5x^7)(2x)^5(32x^5)$
12. $(330 \cdot 10^{11}) \div (11 \cdot 10^2)$	25. $(5x^5)^5(100x^5)(2x^5)^2(80x^2)$
13. $(12 \cdot 10^6) \cdot (5 \cdot 10^6)$	26. $(8x^4)^2(64x^3)(2x^7)^2(16x^6)$

Exponents 6

Solve for x.

1. $16^x = 2^{12}$	9. $9^x \div 9^6 = 9^7$	17. $9^3 = 27^{2x-1}$
2. $32^x = 2^{10}$	10. $4^4 \cdot 4^x \cdot 4^5 = 4^9$	18. $125^{2x-1} = 25^{2x+1}$
3. $3^8 = 9^x$	11. $3^2 \cdot 3^x \cdot 3^4 = 9^6$	19. $16^3 = 2^{2x-4}$
4. $4^x = 8^8$	12. $8^5 \cdot 8^x \div 8^7 = 8^8$	20. $121^{8x} = 1331^{4x+2}$
5. $27^x = 3^6$	13. $11^{16} \div 11^x = 11^9$	21. $4^{3x} = 8^{3x+1}$
6. $16^5 = 2^x$	14. $8^6 \cdot 8^x \div 8^5 = 2^9$	22. $343^{2x+1} = 49^{6x}$
7. $4^x = 64^8$	15. $12^3 \div 2^x = 27$	23. $16^{x+5} = 32^{x-2}$
8. $81^3 = 27^x$	16. $2^x \cdot 2^4 \cdot 2^7 = 4^6$	24. $8^5 = 32^{2x-3}$

Exponents 7

Evaluate. Express as a simplified fraction.

1. 2^{-3}	17. 9^{-3}	33. 1.5^{-3}	49. $(2^{-3})^{-2}$
2. $(-4)^{-3}$	18. 30^{-4}	34. 2.5^{-4}	50. $(9^{-1})^{-1}$
3. 10^{-4}	19. 15^{-2}	35. 1.3^{-2}	51. $(14^{-2})^{-1}$
4. 3^{-5}	20. $(-1)^{-7}$	36. 1.2^{-2}	52. $(.5^{-1})^{-2}$
5. 30^{-3}	21. $(-2)^{-5}$	37. $.6^{-4}$	53. $(1.5^{-1})^{-3}$
6. 80^{-2}	22. 10^{-3}	38. $.3^{-3}$	54. $(3.5^{-1})^{-2}$
7. 11^{-2}	23. 40^{-3}	39. $.25^{-3}$	55. $2^{-2} \div 4^{-1}$
8. 25^{-2}	24. $(-3)^{-3}$	40. $.2^{-3}$	56. $4^{-4} \div 8^{-2}$
9. 1^{-5}	25. 5^{-3}	41. 1.75^{-2}	57. $(1^{-1} + 2^{-1})^{-2}$
10. 6^{-2}	26. 6^{-3}	42. $(-2.5)^{-3}$	58. $(1^{-2} + 2^{-2})^{-2}$
11. 13^{-2}	27. 2^{-4}	43. 1.6^{-2}	59. $(2^{-1} - 3^{-1})^{-1}$
12. 4^{-4}	28. 4^{-3}	44. $.2^{-4}$	60. $(3^{-1} - 3^{-2})^{-1}$
13. 20^{-3}	29. $(-10)^{-4}$	45. $.75^{-3}$	61. $4^{-2} \div 2^{-4}$
14. $(-10)^{-3}$	30. 2^{-6}	46. $(-.4)^{-3}$	62. $4^{-3} \div 2^{-5}$
15. 5^{-4}	31. 8^{-3}	47. $(-1.4)^{-2}$	63. $(3^{-1} - 4^{-1})^{-1}$
16. 12^{-2}	32. 16^{-2}	48. $.25^{-5}$	64. $(2.2^{-1})^{-2}$

Exponents 8

Evaluate.

1. $8^{\frac{1}{3}}$	17. $1331^{\frac{2}{3}}$	33. $16^{\frac{-3}{4}}$
2. $27^{\frac{1}{3}}$	18. $10{,}000^{\frac{3}{4}}$	34. $729^{\frac{-2}{3}}$
3. $1000^{\frac{1}{3}}$	19. $64^{\frac{2}{3}}$	35. $256^{\frac{-1}{2}}$
4. $16^{\frac{1}{4}}$	20. $64^{\frac{3}{2}}$	36. $9^{\frac{-3}{2}}$
5. $4^{\frac{3}{2}}$	21. $625^{\frac{3}{4}}$	37. $8^{\frac{-2}{3}}$
6. $256^{\frac{5}{8}}$	22. $81^{\frac{3}{2}}$	38. $64^{\frac{-2}{3}}$
7. $243^{\frac{3}{5}}$	23. $16^{\frac{5}{4}}$	39. $125^{\frac{-4}{3}}$
8. $128^{\frac{4}{7}}$	24. $(-32)^{\frac{1}{5}}$	40. $243^{\frac{-2}{5}}$
9. $1024^{\frac{3}{10}}$	25. $(-125)^{\frac{1}{3}}$	41. $81^{\frac{-3}{4}}$
10. $32^{\frac{2}{5}}$	26. $(-128)^{\frac{5}{7}}$	42. $(-128)^{\frac{-3}{7}}$
11. $9^{\frac{5}{2}}$	27. $(-729)^{\frac{2}{3}}$	43. $121^{\frac{-3}{2}}$
12. $64^{\frac{5}{3}}$	28. $(-1331)^{\frac{2}{3}}$	44. $125^{\frac{-1}{3}}$
13. $125^{\frac{2}{3}}$	29. $(-125)^{\frac{4}{3}}$	45. $32^{\frac{-3}{5}}$
14. $729^{\frac{1}{3}}$	30. $27^{\frac{1}{3}+\frac{1}{3}}$	46. $(-64)^{\frac{-1}{3}}$
15. $1{,}000{,}000^{\frac{2}{3}}$	31. $16^{\frac{1}{2}-\frac{1}{4}}$	47. $(-32)^{\frac{-4}{5}}$
16. $256^{\frac{1}{4}}$	32. $64^{\frac{1}{2}+\frac{1}{3}}$	48. $(-27)^{\frac{-1}{3}}$

Factorials 1

Simplify.

1. $\dfrac{997!}{996!}$

2. $\dfrac{9!}{4!\ 4!}$

3. $\dfrac{7!\ 7!}{3!\ 10!}$

4. $\dfrac{20!}{18!\ 5!}$

5. $\dfrac{18!}{16!\ 3!}$

6. $\dfrac{40!}{38!\ 3!}$

7. $\dfrac{12!}{8!\ 6!}$

8. $\dfrac{7!\ 8!}{10!}$

9. $\dfrac{15!}{6!\ 11!}$

10. $\dfrac{5!\ 6!}{10!}$

11. $\dfrac{8!}{4!\ 5!}$

12. $\dfrac{100!}{2!\ 99!} + \dfrac{12!}{9!}$

13. $\dfrac{13!}{12!} + \dfrac{11!}{10!}$

14. $\dfrac{8!}{10!} + \dfrac{4!}{6!}$

15. $\dfrac{7!}{9!} + \dfrac{2!}{4!}$

16. $\dfrac{3!}{5!} - \dfrac{3!}{6!}$

17. $\dfrac{2!}{5!} + \dfrac{7!}{10!}$

18. $\dfrac{(30-10)!}{17!\ 2!\ 3!}$

19. $\dfrac{28!\ 31!}{27!\ 32!}$

20. $\dfrac{27!\ 38!}{25!\ 40!}$

21. $\dfrac{(9-2)!\ 8!}{3!\ 5!\ 7!}$

22. $\dfrac{15!}{7!\ 9!}$

Factorials 2

Find the number of rightmost zeros when written in standard form.	*Find the number of specified factors of the given factorial.*
1. 13!	12. 10! has how many factors of 6?
2. 20!	13. 10! has how many factors of 12?
3. 30!	14. 15! has how many factors of 9?
4. 36!	15. 8! has how many factors of 2?
5. 41!	16. 30! has how many factors of 9?
6. 48!	17. 10! has how many factors of 8?
7. 63!	18. 10! has how many factors of 20?
8. 77!	19. 14! has how many factors of 14?
9. 150!	20. 25! has how many factors of 15?
10. 250!	21. 18! has how many factors of 6?
11. 500!	22. 100! has how many factors of 2?

Factorials 3

Answer as indicated.

1. Find the ones digit of 5! + 6! + . . . + 12!.	7. Evaluate $\dfrac{(n + 2)!}{(n - 2)!}$ for n = 9.
2. Find the least whole number n such that n! + (n+1)! is a multiple of 10.	8. How many odd factors does 11! have?
3. Find the greatest odd factor of 6!. Find the greatest odd factor of 7!	9. Find the greatest power of 6 as a factor of 12!. 12! can be divided by 6 how many times with no remainder?
4. Find the greatest value of n if 2^n is a factor of 100!.	10. Find the greatest value of n if 3^n is a factor of 90!.
5. When 20! is written in base 3, how many rightmost zeros will it have?	11. When 30! is written in base 5, how many rightmost zeros will it have?
6. a = the least whole number such that a! is a multiple of 10 b = the only whole number such that b! is the product of two primes c = the only whole number such that c! is prime Find a + b + c.	12. a = the least whole number such that a! is a multiple of 8 b = the least whole number such that b! has 4 prime factors c = the least whole number such that c! is a multiple of 9 Find abc.

Factorials 4

Write the prime factorization. Do each without referring to another answer.	Solve for n.
1. $5!$	11. $\dfrac{11!}{n!} = 990$
2. $9!$	12. $\dfrac{10!}{(n-1)!} = 720$
3. $10!$	13. $\dfrac{12!}{(n+1)!} = 1320$
4. $13!$	14. $\dfrac{n!}{4!\,(n-3)!} = \dfrac{n!}{5!\,(n-4)!}$
5. $7!$	15. $\dfrac{14!}{7!\,(n-4)!} = \dfrac{14!}{8!\,(n-5)!}$
6. $11!$	16. $\dfrac{8!}{6! + 6!} = 7n$
7. $6!$	17. $\dfrac{10!}{7! + 7!} = 20n$
8. $8!$	18. $\dfrac{n!}{6!\,(n-3)!} = \dfrac{n!}{4!\,(n-2)!}$
9. $12!$	19. $\dfrac{n!}{7!\,(n-5)!} = \dfrac{n!}{9!\,(n-6)!}$
10. $14!$	20. $\dfrac{7!}{4! + 4!} = 21n$

Factors 1

List all factors in ascending order, working in from both sides.

1. 48	12. 360
2. 50	13. 420
3. 64	14. 550
4. 72	15. 640
5. 80	16. 750
6. 93	17. 888
7. 100	18. 980
8. 120	19. 1001
9. 150	20. 1100
10. 196	21. 1250
11. 275	22. 2000

Factors 2

Find the missing number by examining and matching factors. Do not multiply.

1. $11 \cdot 86 = 22 \cdot ?$

2. $12 \cdot 95 = 60 \cdot ?$

3. $45 \cdot 16 = 90 \cdot ?$

4. $32 \cdot 10 = 20 \cdot ?$

5. $40 \cdot 25 = 20 \cdot ?$

6. $36 \cdot 22 = 18 \cdot ?$

7. $51 \cdot 62 = 34 \cdot ?$

8. $24 \cdot 30 = 80 \cdot ?$

9. $54 \cdot 16 = 72 \cdot ?$

10. $22 \cdot 85 = 11 \cdot ?$

11. $14 \cdot 15 = 10 \cdot ?$

12. $35 \cdot 44 = 55 \cdot ?$

13. $14 \cdot 51 \cdot 48 = 34 \cdot 42 \cdot ?$

14. $96 \cdot 95 \cdot 42 = 56 \cdot 60 \cdot ?$

15. $25 \cdot 12 \cdot 26 = 5 \cdot 39 \cdot ?$

16. $12 \cdot 14 \cdot 85 = 20 \cdot 21 \cdot ?$

17. $125 \cdot 14 = 50 \cdot ?$

18. $21 \cdot 18 \cdot 22 = 14 \cdot 99 \cdot ?$

19. $121 \cdot 27 \cdot 35 = 33 \cdot 45 \cdot ?$

20. $72 \cdot 45 \cdot 15 = 24 \cdot 25 \cdot ?$

21. $15 \cdot 49 \cdot 65 = 21 \cdot 25 \cdot ?$

22. $34 \cdot 10 \cdot 30 = 50 \cdot 51 \cdot ?$

Factors 3

Write each counting number from 1 to 84 in the column labeled by the counting number's number of factors.

1	2	3	4	5	6	7	8	9	10	11	12

Factors 4

Find the number of rightmost zeros of each product without multiplying.

1. $2^8 \cdot 5^7$	12. $30^4 \cdot 45^2 \cdot 55^2$
2. $2^7 \cdot 25^3$	13. $16^3 \cdot 125^3$
3. $35!$	14. $75!$
4. $8^3 \cdot 25^4$	15. $5! \cdot 5^3 \cdot 500$
5. $4 \cdot 5 \cdot 8 \cdot 10 \cdot 14 \cdot 15$	16. $7! \cdot 8! \cdot 9! \cdot 10! \cdot 11!$
6. $320 \cdot 1250$	17. $96 \cdot 675$
7. $420 \cdot 7500$	18. $60 \cdot 65 \cdot 70 \cdot 75 \cdot 80$
8. $6 \cdot 12 \cdot 16 \cdot 20 \cdot 75$	19. $60^4 \cdot 15^3$
9. $105 \cdot 625$	20. $8! \cdot 35^3$
10. $25 \cdot 30 \cdot 35 \cdot 40 \cdot 45 \cdot 50$	21. $100 \cdot 200 \cdot 300 \cdot 400 \cdot 500$
11. $225 \cdot 128$	22. $2! \cdot 3! \cdot 4! \cdot 5! \cdot 6!$

Factors 5

Use multiplication principle to find the number of:	$2^3 \cdot 5^6 \cdot 7^4$	$2^3 \cdot 3^4 \cdot 5^2 \cdot 7^2 \cdot 11$
factors		
prime factors		
composite factors		
even factors		
odd factors		
perfect square factors		
perfect cube factors		
factors with ones digit 0		
factors with ones digit 5		
multiple of 49 factors		

	$2^4 \cdot 3^3 \cdot 5 \cdot 13^2$	$2^2 \cdot 3^5 \cdot 5^3 \cdot 7^3 \cdot 17^2$
factors		
prime factors		
composite factors		
even factors		
odd factors		
perfect square factors		
perfect cube factors		
factors with ones digit 0		
factors with ones digit 5		
multiple of 27 factors		
multiple of 36 factors		

Factors 6

Use multiplication principle to find the number of:	$6^3 \cdot 35^2 \cdot 22^4$	$8^2 \cdot 15^3 \cdot 49^2 \cdot 121$
factors		
composite factors		
even factors		
odd factors		
perfect square factors		
perfect cube factors		
factors with ones digit 0		
factors with ones digit 5		
multiple of 33 factors		

	$10^4 \cdot 21^3 \cdot 30^2$	$6^4 \cdot 14 \cdot 35^2 \cdot 64$
factors		
prime factors		
composite factors		
even factors		
odd factors		
perfect square factors		
perfect cube factors		
factors with ones digit 0		
factors with ones digit 5		
multiple of 21 factors		

Factors 7

Given factors of a number, find all other factors conclusively determinable.	Find the number of digits when written in standard form. Do not multiply.
1. 1, 3, 7, 11	14. $2^{10} \cdot 25^6$
2. 1, 2, 11, 17	15. $2^3 \cdot 5^6 \cdot 10^2$
3. 1, 5, 11, 19	16. $2^5 \cdot 5^3 \cdot 20^2$
4. 1, 2, 3, 4, 5	17. $2^3 \cdot 5^6 \cdot 6^3$
5. 1, 3, 4, 5	18. $4^4 \cdot 5^4 \cdot 10^2$
6. 1, 2, 3, 5, 7	19. $4^3 \cdot 5^7 \cdot 8^2$
7. 1, 3, 9, 10	20. $10^2 \cdot 15^2 \cdot 20^2$
8. 1, 2, 4, 9, 11	21. $2^6 \cdot 5^3 \cdot 30^2$
9. 1, 3, 5, 9	22. $4^3 \cdot 8^2 \cdot 25^4$
10. 1, 7, 9, 11	23. $20^2 \cdot 25^3 \cdot 30^2$
11. 1, 7, 8, 11	24. $15^2 \cdot 20^3 \cdot 35^2$
12. 1, 5, 6, 23	25. $6^3 \cdot 8^3 \cdot 25^5$
13. 1, 6, 10, 12	26. $10^4 \cdot 25^3 \cdot 40^3$

Factors 8

Answer as indicated.

1. Find the sum of all two-digit numbers with three factors.	9. Find the greatest 5-digit even number such that each pair of adjacent digits is relatively prime (GCF = 1).
2. Find the sum of the greatest 2-digit number and the least 3-digit number, each with an odd number of factors.	10. Find the sum of the greatest 2-digit number and the least 3-digit number, each with exactly three factors.
3. Find the sum of the greatest 3-digit number and the least 4-digit number, each with exactly three factors.	11. Find the sum of the greatest 3-digit number and the least 4-digit number, each with an odd number of factors.
4. If ABCABC is a 6-digit number, find the greatest prime factor that ABCABC must have.	12. Find the sum of all the two-digit factors of 140.
5. Find the sum of the two-digit numbers with 9 or 10 factors.	13. Find the sum of all numbers less than 1000 with exactly 5 factors.
6. Find the least natural number with exactly 5 factors.	14. Find the least natural number with exactly 10 factors.
7. How many two-digit numbers do not have a factor of 17?	15. Find the sum of all the two-digit factors of 200.
8. Find the greatest 6-digit number divisible by 4 such that each pair of adjacent digits has a GCF of 2.	16. Find the least 6-digit number divisible by 22 such that each pair of adjacent digits is relatively prime.

Fictitious Operations 1

Evaluate the unary operations, following the definitions.

Define operation @ by @x = x^2 + 3.
For example, @4 = 19.

1. @7

2. @(–9)

3. @(.5)

Define operation [ ] by [x] = x^2 – 1.
For example, [3] = 8.

16. [4]

17. [–6]

18. [2] + [5]

Define operation ∂ by ∂x = x^3 – 5.
For example, ∂5 = 120.

4. ∂(2)

5. ∂(–4)

6. ∂(.1)

7. ∂(11)

Define operation ≈ ≈ by ≈x≈ = 2x + 11.
For example, ≈4≈ = 19.

19. ≈2≈

20. ≈(–10)≈

21. ≈(.5)≈

22. ≈(17.5)≈

Define operation ~ ~ by ~x~ = 5x + 2.
For example, ~4~ = 22.

8. ~2~

9. ~(–5)~

10. ~1.2~

11. ~(–3.1)~

Define operation ‡ ‡ by ‡x‡ = $2x^2$ – 2.
For example, ‡3‡ = 16.

23. ‡(.5)‡

24. ‡(–5)‡

25. ‡4‡

26. ‡11‡

Define operation Ω by Ωx = (.5)x if x is
even and Ωx = 2x if x is odd.
For example, Ω4 = 2 and Ω9 = 18.

12. (Ω5)(Ω10)

13. Ω3 + Ω5

14. Ω20 ÷ Ω8

15. Ω14 – Ω7

Define operation ¥ by ¥x = 2x if x is odd
and ¥x = 3x if x is even. For example,
¥4 = 12 and ¥3 = 6.

27. ¥7 – ¥10

28. ¥8 ÷ ¥5

29. (¥11)(¥2)

30. ¥3 + ¥12

Fictitious Operations 2

Evaluate the binary operations, following the definitions.

Define operation @ by a@b = a² + 3b.
For example, 3@4 = 9 + 12 = 21.

1.   5@7

2.   4@(−9)

3.   6@(.5)

Define operation [ ] by [x, y] = |y| − |y − x|.
For example, [3, −4] = 4 − 7 = −3.

17.   [−2, −5]

18.   [6, −3]

19.   [−7, 0]

Define operation ß by x ß y = 2xy + x.
For example, 2 ß 3 = 12 + 2 = 14.

4.   3 ß 8

5.   −5 ß 4

6.   6 ß (−5)

7.   −8 ß 2

Define operation Δ by a Δ b = 8b ÷ a.
For example, 4 Δ 2 = 4.

20.   4 Δ 5

21.   3 Δ 6

22.   6 Δ 3

23.   5 Δ 4

Define operation § by x § y = x − y + 2.
For example, 7 § 5 = 2 + 2 = 4.

8.   5 § 8

9.   6 § 6

10.   4 § 11

11.   −3 § (−9)

Define operation ¶ by x ¶ y = y − x + 3.
For example, 9 ¶ 2 = −7 + 3 = −4.

24.   8 ¶ 5

25.   12 ¶ 12

26.   4 ¶ 10.5

27.   12 ¶ 6

Define operation ∞ by x ∞ y = −xy − 5.
For example, 2 ∞ 3 = −6 − 5 = −11.

12.   8 ∞ −7

13.   −4 ∞ −9

14.   −3 ∞ 6

15. $\frac{3}{4}$ ∞ $\frac{4}{3}$

16. $\frac{12}{5}$ ∞ $\frac{5}{6}$

Define operation • by a • b = ab + a.
For example, 4 • 2 = 8 + 4 = 12.

28.   3 • 4

29. $\frac{7}{5}$ • $\frac{5}{7}$

30. $\frac{9}{8}$ • 16

31.   −5 • 7

32.   8 • 9

Fictitious Operations 3

Evaluate the nested or grouped operations by following the definitions.

Define operation @ by @x = x^2 – 5.
For example, @5 = 20.

1. @(@1)

2. @(@2)

3. @(@4)

Define operation ß by x ß y = 3xy – y.
For example, 5 ß 6 = 84.

4. (2 ß 3) ß 4

5. 2 ß (3 ß 4)

6. (.5 ß 6) ß 5

7. –1 ß (7 ß 3)

Define operation ~ ~ by ~x~ = 4x + 3.
For example, ~4~ = 16 + 3 = 19.

8. ~(~5~)~

9. ~(~.5~)~

10. ~(~3~)~

11. ~(~1.5~)~

Define operation ∞ by x ∞ y = –2xy + y.
For example, 2 ∞ 3 = –12 + 3 = –9.

12. (3 ∞ –4) ∞ 5

13. (–5 ∞ 2) ∞ 3

14. 6 ∞ (–2 ∞ 4)

15. –2 ∞ (4 ∞ 5)

16. (–1 ∞ 9) ∞ 0

17. –3 ∞ (–2 ∞ 7)

Define operation [ ] by [x] = $2x^2$ + 1.
For example, [5] = 50 + 1 = 51.

18. [[2]]

19. [[–1]]

20. [[4]]

Define operation ∆ by a ∆ b = 10b ÷ a.
For example, 4 ∆ 2 = 5.

21. (2 ∆ 5) ∆ 5

22. 2 ∆ (3 ∆ 6)

23. (6 ∆ 3) ∆ 10

24. (4 ∆ 8) ∆ 5

Define operation ‡ ‡ by ‡x‡ = $-x^2$ + 4.
For example, ‡7‡ = –45.

25. ‡(‡2‡)‡

26. ‡(‡(–1)‡)‡

27. ‡(‡–3‡)‡

28. ‡(‡4‡)‡

Define operation • by a • b = ab – |2a|.
For example, 7 • 8 = 56 – 14 = 42.

29. (–2 • 5) • 3

30. (8 • 2) • 4

31. –3 • (–2 • 3)

32. –6 • (4 • 5)

33. (–1 • 6) • 5

34. –4 • (–3 • 5)

Fictitious Operations 4

Evaluate the backwards operations by solving the equations.

Define operation @ by $x@y = x^3 - 3 + 2y$. For example, $2@12 = 8 - 3 + 24 = 29$. 1. $3@a = 50$ 2. $(-1)@z = 32$ 3. $w@5 = 71$	Define operation [ ] by $[x, y] = y^2 - 3x + 4$. For example, $[3, 2] = 4 - 9 + 4 = -1$. 17. $[w, 5] = 2$ 18. $[a, -3] = 9$ 19. $[2, c] = 34$
Define operation ß by $x ß y = 3y + 4x - 2$. For example, $3 ß 2 = 6 + 12 - 2 = 16$. 4. $4 ß w = 56$ 5. $c ß 3 = 47$ 6. $5 ß e = 72$ 7. $c ß 5 = 69$	Define operation Δ by $a Δ b = 30b ÷ a + 1$. For example, $10 Δ 2 = 6 + 1 = 7$. 20. $y Δ 4 = 13$ 21. $5 Δ w = 9$ 22. $6 Δ x = 16$ 23. $x Δ 8 = 49$
Define operation ~~ by $x{\sim}{\sim}y = 5y - 7x$. For example, $4{\sim}{\sim}3 = 15 - 28 = -13$. 8. $c{\sim}{\sim}8 = 33$ 9. $-3{\sim}{\sim}w = 46$ 10. $b{\sim}{\sim}6.2 = 10$ 11. $-9{\sim}{\sim}z = 98$	Define operation ‡ by $a‡b = 2a^2 - 1 + b$. For example, $3‡5 = 18 - 1 + 5 = 22$. 24. $6‡z = 90$ 25. $w‡(-6) = 43$ 26. $-4‡x = 33$ 27. $c‡11 = 42$
Define operation μ by $μx = 3x$ if x is even and $μx = 5x$ if x is odd. For example, $μ4 = 12$ and $μ9 = 45$. 12. $(μw)(μ10) = 450$ 13. $μ5 + μy = 43$ 14. $μ(μx) = 18$ 15. $μc + μ8 = 60$ 16. $(μ5)(μw) = 125$	Define operation £ by $£x = 2x$ if x is prime and $£x = 4x$ if x is composite. For example, $£7 = 14$ and $£6 = 24$. $£1$ is undefined. 28. $£13 - £y = -14$ 29. $£w ÷ £5 = 8.8$ 30. $£(£x) = 16$ 31. $(£3)(£w) = 96$ 32. $£6 + £y = 30$

Fractional Parts 1

Find the fractional part by forming the "is/of" fraction and simplifying.

1. 15 seconds is what fractional part of 15 hours?

2. 2 cups is what fractional part of 3 gallons?

3. 440 is what fractional part of 1870?

4. What fractional part of one and one-third hours is two minutes?

5. Commercials aired for 16 minutes on an hour TV show. What fractional part of the entire show were the commercials?

6. If 12 scripts were read and 18 were unread, what fraction of the scripts were unread?

7. 6 inches is what fractional part of 2 yards?

8. The days of January, February, and March are what fractional part of a non-leap year?

9. 5 pints is what fractional part of 10 gallons?

10. What fraction of 20 hours is 10 seconds?

11. 1020 is what fractional part of 8500?

12. What fraction of one and one-quarter hours is half a minute?

13. Commercials played for 6 minutes on a half hour radio show. For what fractional part of the show were commercials not shown?

14. A basketball team won 6 games and lost 30. What fraction of its games did the team win?

15. In a class of 40 students, 24 are girls. What fraction of the students are boys?

16. In a jar of 12 red, 18 blue, 25 green and 50 purple marbles, what fraction of the marbles are not red?

Fractional Parts 2

Answer as indicated, drawing vertical trees with discards to the left.

1. Twenty-four apples were on a tree. One third fell off and 1/4 of the remainder were picked. What fraction of the original apples were left?	5. Eighteen apples are on a tree. One third fall off. Of those left, someone picks 1/3. What fraction of the original apples are still on the tree?
2. A carpenter used 1/3 of his lumber for one project and 3/5 of what was left for another project. If he started with 30 units of lumber, what fraction of the original lumber remained?	6. Abby baked 80 cookies and threw out 1/4 that crumbled. Of those whole, she gave 1/6 to her sisters. Of those then left, she wrapped and froze 3/5. What fraction of the original cookies remained?
3. Dan had 96 marbles. He lost 1/4. He gave 1/6 of those left to a friend and 2/5 of those then remaining to another friend. What fraction of the original marbles did Dan then have?	7. Jon had 100 marbles. He lost 1/5. Of those left, he gave 1/5 to a friend, 1/4 to a brother, and kept the rest. What fraction of the marbles did Jon keep?
4. Forty-eight lemons were on a tree. One third fell off and 1/8 of the remainder were picked. What fraction of the original apples were left?	8. If a tree has 36 bananas, 1/3 are picked, and then 1/8 of those left are picked, what fraction of the bananas were picked?

Fractional Parts 3

Answer as indicated, drawing vertical trees with discards to the left.

1.

A baker gave
1/2 of his cookies
to a charity. After he
boxed 1/5 of those
left, 40 cookies
remained. Find the
number of cookies
at the start.

5.

Ted saved 12% of
his marbles and
gave 24% to a
friend. Of those left,
he lost 25%. If 60
marbles remained,
how many did Ted
have originally?

2. Jon gave 30% of his cards to his brother, 10% to his sister, and 20% of those left to a friend. If 24 cards remained, how many did Jon have to start?

6.

Dana ate 1/4 of her
candies and froze 1/3
of those left. She then
packed 1/3 of those
remaining, leaving 20.
How many candies
did Dana have
originally?

3.

One third of the
students in a class
went to the movies,
1/2 of those left went
to the game, and the
remaining 10 stayed
home. How many
students were in the
class?

7.

Of the apples on a
tree, 20% fell off. Of
those left, 15% were
boxed. Of those left,
25% were discarded.
How many were on
the tree to start if 102
apples remained?

4.

Hal boxed 1/6 of his
books and gave away
1/4. Of those left,
1/7 went to charity. Of
those left, he sold 1/2,
leaving 30. What was
his original number of
books?

8. One-sixth of a group ate pizza only, 1/4 ate sandwiches only, and 1/3 of the others ate soup only. If 24 people had no food, how many people were in the group?

Fractional Parts 4

Find the fractional part.

1. 1.04 is what fractional part of 4.16?

2. 4.9 is what fractional part of 12.25?

3. 5.2 is what fractional part of 21.6?

4. $10\frac{2}{5}$ is what fractional part of $12\frac{2}{5}$?

5. $7\frac{1}{9}$ is what fractional part of $14\frac{2}{3}$?

6. 30% is what fractional part of 75%?

7. 20% is what fractional part of 85%?

8. $6\frac{1}{4}$ is what fractional part of $16\frac{2}{3}$?

9. $6\frac{5}{6}$ is what fractional part of $13\frac{2}{3}$?

10. 3.25 is what fractional part of 15.25?

11. $1.\overline{5}$ is what fractional part of $1.\overline{7}$?

12. 15% is what fractional part of 35%?

13. $11\frac{1}{9}$ is what fractional part of $15\frac{5}{6}$?

14. 10.5 is what fractional part of 9.1?

15. 8.6 is what fractional part of 60.2?

16. $8\frac{5}{9}$ is what fractional part of $17\frac{3}{5}$?

17. 12% is what fractional part of 66%?

18. 20.1 is what fractional part of 21.3?

19. $6\frac{3}{4}$ is what fractional part of $12\frac{3}{8}$?

20. 3.12 is what fractional part of 5.16?

21. $6\frac{3}{7}$ is what fractional part of $7\frac{13}{14}$?

22. $3.\overline{1}$ is what fractional part of $4.\overline{4}$?

Fractions 1

Simplify.

1. $\dfrac{150}{360}$	12. $\dfrac{112}{224}$	23. $\dfrac{147}{196}$
2. $\dfrac{630}{840}$	13. $\dfrac{135}{495}$	24. $\dfrac{510}{570}$
3. $\dfrac{105}{420}$	14. $\dfrac{315}{385}$	25. $\dfrac{280}{175}$
4. $\dfrac{280}{180}$	15. $\dfrac{168}{252}$	26. $\dfrac{352}{374}$
5. $\dfrac{130}{650}$	16. $\dfrac{720}{990}$	27. $\dfrac{126}{423}$
6. $\dfrac{112}{168}$	17. $\dfrac{105}{231}$	28. $\dfrac{504}{630}$
7. $\dfrac{770}{630}$	18. $\dfrac{462}{168}$	29. $\dfrac{128}{512}$
8. $\dfrac{225}{900}$	19. $\dfrac{560}{640}$	30. $\dfrac{325}{875}$
9. $\dfrac{810}{720}$	20. $\dfrac{126}{612}$	31. $\dfrac{594}{154}$
10. $\dfrac{225}{275}$	21. $\dfrac{484}{363}$	32. $\dfrac{300}{375}$
11. $\dfrac{385}{231}$	22. $\dfrac{147}{210}$	33. $\dfrac{198}{594}$

Fractions 2

Multiply, simplifying first.

1.

$$\frac{13}{14} \times \frac{21}{25} \times \frac{15}{39} \times \frac{5}{6}$$

8.

$$\frac{11}{36} \times \frac{42}{22} \times \frac{13}{39} \times \frac{6}{7}$$

2.

$$\frac{45}{42} \times \frac{10}{15} \times \frac{20}{50} \times \frac{1}{4}$$

9.

$$\frac{29}{78} \times \frac{34}{15} \times \frac{39}{17} \times \frac{16}{58}$$

3.

$$\frac{21}{12} \times \frac{36}{35} \times \frac{15}{28} \times \frac{8}{9}$$

10.

$$\frac{5}{9} \times \frac{27}{26} \times \frac{63}{84} \times \frac{26}{15} \times \frac{900}{225}$$

4.

$$\frac{15}{14} \times \frac{11}{20} \times \frac{16}{55} \times \frac{70}{80} \times \frac{1}{3}$$

11.

$$\frac{2}{5} \times \frac{70}{14} \times \frac{10}{21} \times \frac{15}{35} \times \frac{49}{24}$$

5.

$$\frac{16}{15} \times \frac{55}{48} \times \frac{30}{25} \times \frac{30}{11} \times \frac{6}{5}$$

12.

$$\frac{16}{55} \times \frac{15}{14} \times \frac{11}{20} \times \frac{14}{24} \times \frac{5}{2}$$

6.

$$\frac{22}{12} \times \frac{16}{50} \times \frac{25}{48} \times \frac{33}{10} \times \frac{120}{121}$$

13.

$$\frac{21}{20} \times \frac{15}{35} \times \frac{12}{27} \times \frac{16}{10} \times \frac{5}{4}$$

7.

$$\frac{45}{32} \times \frac{56}{14} \times \frac{48}{63} \times \frac{70}{50} \times \frac{2}{5}$$

14.

$$\frac{98}{15} \times \frac{45}{56} \times \frac{36}{49} \times \frac{42}{54}$$

Fractions 3

Add or subtract as indicated.

1.
$$\frac{7}{24} + \frac{3}{40}$$

8.
$$\frac{5}{21} + \frac{13}{14} + \frac{1}{9}$$

2.
$$\frac{11}{50} + \frac{13}{70}$$

9.
$$\frac{1}{15} + \frac{5}{12} + \frac{7}{4}$$

3.
$$\frac{7}{60} - \frac{3}{40}$$

10.
$$\frac{2}{21} + \frac{5}{24} + \frac{3}{28}$$

4.
$$\frac{3}{76} + \frac{1}{20}$$

11.
$$\frac{3}{20} - \frac{2}{15} + \frac{1}{4}$$

5.
$$\frac{2}{45} + \frac{7}{18}$$

12.
$$\frac{5}{22} + \frac{3}{33} - \frac{4}{55}$$

6.
$$\frac{9}{40} + \frac{5}{56}$$

13.
$$\frac{5}{18} + \frac{5}{24} - \frac{1}{36}$$

7.
$$\frac{5}{36} + \frac{7}{60}$$

14.
$$\frac{4}{21} + \frac{2}{15} - \frac{2}{35}$$

Fractions 4

Operate and simplify.

1.
$$\dfrac{\dfrac{1}{4} + \dfrac{3}{8} + \dfrac{5}{12}}{\dfrac{5}{12}}$$

2.
$$\dfrac{\dfrac{3}{7} + \dfrac{1}{5}}{\dfrac{11}{40}}$$

3.
$$\dfrac{\dfrac{5}{6}}{\dfrac{1}{2}} + \dfrac{1}{3}$$

4.
$$\dfrac{\dfrac{5}{6} - \dfrac{1}{3}}{\dfrac{2}{9} + \dfrac{1}{6}}$$

5.
$$\dfrac{\dfrac{3}{16} + 6}{9 - \dfrac{3}{8}}$$

6.
$$\dfrac{\dfrac{3}{11} + 2}{2 - \dfrac{3}{4}}$$

7.
$$\dfrac{\dfrac{2}{5} + \dfrac{5}{6} + \dfrac{7}{15}}{\dfrac{17}{60}}$$

8.
$$\dfrac{\dfrac{4}{5} + \dfrac{3}{10}}{\dfrac{33}{20}}$$

9.
$$\dfrac{\dfrac{9}{10} + \dfrac{5}{6}}{\dfrac{9}{10} - \dfrac{5}{6}}$$

10.
$$\dfrac{\dfrac{5}{6} + \dfrac{3}{4}}{\dfrac{5}{6} - \dfrac{3}{4}}$$

11.
$$\dfrac{\dfrac{5}{6} \times \dfrac{2}{3}}{\dfrac{2}{3} \div \dfrac{5}{6}}$$

12.
$$\dfrac{\dfrac{5}{6} + \dfrac{2}{3}}{\dfrac{2}{3} \times \dfrac{5}{6}}$$

13.
$$\dfrac{\dfrac{3}{4} \times \dfrac{2}{3}}{\dfrac{2}{3} + \dfrac{3}{4}}$$

14.
$$\dfrac{\dfrac{15}{16} \times \dfrac{4}{5}}{\dfrac{5}{12} \div \dfrac{5}{6}}$$

Fractions 5

Operate and simplify. Answer as a mixed number.

1. $5 + \dfrac{1}{3 + \dfrac{1}{2}}$

2. $2 + \dfrac{1}{1 + \dfrac{1}{6}}$

3. $1 + \dfrac{1}{2 + \dfrac{1}{4}}$

4. $8 + \dfrac{1}{3 + \dfrac{1}{3}}$

5. $6 + \dfrac{1}{3 + \dfrac{1}{5}}$

6. $7 - \dfrac{1}{1 + \dfrac{1}{1 + \dfrac{1}{7}}}$

7. $2 + \dfrac{1}{4 - \dfrac{1}{2 + \dfrac{1}{3}}}$

8. $5 - \dfrac{1}{1 + \dfrac{2}{3 + \dfrac{1}{4}}}$

9. $6 - \dfrac{3}{5 - \dfrac{7}{9}}$

10. $3 - \dfrac{2}{4 - \dfrac{1}{8}}$

11. $5 + \dfrac{4}{3 - \dfrac{1}{5}}$

12. $4 - \dfrac{2}{4 - \dfrac{7}{11}}$

13. $5 - \dfrac{2}{3 - \dfrac{4}{5}}$

14. $3 + \dfrac{1}{5 - \dfrac{1}{4 - \dfrac{1}{5}}}$

15. $4 - \dfrac{1}{2 + \dfrac{1}{3 - \dfrac{1}{4}}}$

16. $1 + \dfrac{1}{3 + \dfrac{1}{3 + \dfrac{1}{3}}}$

Fractions 6

Rewrite in ascending order using multiple techniques: mental math, decimal conversion, common denominator, versus 1/2, missing part (same N or D), and pairwise comparison.

1. $\frac{4}{7}$ $\frac{5}{9}$ $\frac{3}{5}$ $\frac{1}{2}$

2. $\frac{13}{20}$ $\frac{20}{33}$ $\frac{11}{23}$ $\frac{1}{2}$

3. $\frac{12}{25}$ $\frac{9}{17}$ $\frac{9}{19}$ $\frac{11}{21}$

4. $\frac{7}{10}$ $\frac{11}{20}$ $\frac{14}{25}$ $\frac{23}{50}$

5. $\frac{4}{5}$ $\frac{2}{3}$ $\frac{6}{7}$ $\frac{7}{9}$

6. $\frac{7}{12}$ $\frac{27}{48}$ $\frac{13}{24}$ $\frac{5}{8}$

7. $\frac{8}{13}$ $\frac{11}{15}$ $\frac{8}{11}$ $\frac{2}{3}$

8. $\frac{3}{20}$ $\frac{1}{9}$ $\frac{4}{25}$ $\frac{7}{40}$

9. $\frac{5}{6}$ $\frac{23}{30}$ $\frac{4}{5}$ $\frac{11}{15}$

10. $\frac{11}{17}$ $\frac{11}{18}$ $\frac{6}{11}$ $\frac{15}{31}$

11. $\frac{32}{33}$ $\frac{30}{31}$ $\frac{97}{99}$ $\frac{59}{62}$

12. $\frac{32}{95}$ $\frac{31}{99}$ $\frac{30}{91}$ $\frac{31}{94}$

MAVA Math: Enhanced Skills Copyright © 2015 Marla Weiss

Fractions 7

Operate using fractions. Answer as a simplified fraction.

1. 10% of 40% of $\frac{3}{4}$ of 1.25	9. 52% of 25% of 22 + $\frac{2}{7}$ of 63%
2. $\frac{2}{3}$ of 25% of 5% of 5.4	10. 18.75% of 11.2 − $\frac{4}{9}$ of 12% of 22.5
3. $\frac{1}{3}$ of 80% of 2.4 + $\frac{1}{2}$ of 6% of 1.$\overline{3}$	11. $\frac{5}{9}$ of 50% of 6.3 + $\frac{5}{6}$ of 24%
4. $\frac{4}{5}$ of 60% of 2.25 + $\frac{5}{3}$ of 8% of 5.1	12. 6.$\overline{6}$% of 62.5% of 0.1$\overline{6}$ of 0.48
5. $\frac{5}{9}$ of 5% of 45% of $\frac{3}{4}$ of 20.8	13. 0.25 of 12% of $\frac{16}{27}$ of 8% of 4.5
6. .375 of 40% of $\frac{2}{3}$ of 62.5% of 3.2	14. 2.2 of 15% of 6.4 of $\frac{2}{11}$ of 15.625%
7. $\frac{4}{5}$ of 65% + $\frac{1}{3}$ of 5.4 of 80%	15. $\frac{3}{5}$ of 4.5% of $\frac{2}{9}$ of .125 of 4.$\overline{4}$
8. 30% of 62.5% of 4.8 + .125 of 8.8	16. 5.8$\overline{3}$ of 87.5% of $\frac{20}{49}$ of 10.5%

Fractions 8

Operate on natural numbers. Simplify before multiplying. Answer as a simplified fraction.

1. Divide the product of the 6 least composite numbers by the product of the next 6 composites.	7. Divide the product of the 6 least 2-digit composite numbers by the product of the next 5 composites.
2. Divide the product of the first 5 2-digit composites greater than 19 by the product of the first 5 primes.	8. Divide the product of the 2-digit composites with tens digit 4 by the product of the first 7 squares.
3. Divide the product of the 2-digit composites with ones digit 9 by the product of the 1st 4 numbers with ones digit 3.	9. Divide the product of the 2-digit composites with ones digit 3 by the product of the 1st three 2-digit numbers with ones digit 1.
4. Divide the product of the 2-digit composites with ones digit 1 by the product of the first 9 odd numbers.	10. Divide the product of the 2-digit composites with ones digit 1 by the product of the 1st 3 numbers with ones digit 7.
5. Divide the product of the first 6 composites greater than 29 by the product of the first 7 squares.	11. Divide the product of the first 6 composites greater than 39 by the product of the first 6 cubes.
6. Divide the product of the first 7 multiples of 4 by the product of the first 7 multiples of 6. Take the 7th root.	12. Divide the product of the first 6 multiples of 6 by the product of the first 6 multiples of 15. Take the 6th root.

Fractions 9

Answer by making a chart.

1. A flower garden is 1/8 pansies, 1/14 lilies, 1/2 roses, and 1/6 gardenias, while the remaining 46 flowers are irises. How many of the flowers are lilies?

4. A flower garden is 1/5 pansies, 2/15 lilies, 1/4 roses, and 3/20 gardenias, while the remaining 48 flowers are irises. How many of the flowers are pansies?

2. A vegetable garden is 1/3 onions, 1/10 carrots, 2/7 radishes, and 2/15 peppers. The other 93 plants are lettuces. How many of each type of vegetable are in the garden?

5. A vegetable garden is 1/8 onions, 1/12 carrots, 2/5 radishes, and 4/15 peppers. The other 45 plants are lettuces. How many of each type of vegetable are in the garden?

3. An herb garden is 1/12 parsley, 3/14 sage, 4/21 rosemary, and 5/18 thyme. The remaining 177 herb plants are dill. How many of the herbs are thyme?

6. An herb garden is 1/15 parsley, 1/3 sage, 3/10 rosemary, and 7/50 thyme. The remaining 96 herb plants are dill. Find the total number of herb plants in the garden.

Fractions 10

Find the number of unique values of a/b, where a and b are distinct members of the given set.

1. {2, 4, 8}

2. {2, 6, 9}

3. {3, 9, 27}

4. {2, 8, 32}

5. {2, 4, 6}

6. {2, 5, 6, 15}

7. {2, 4, 8, 16}

8. {3, 6, 12, 24}

9. {2, 6, 18, 54}

10. {2, 3, 8, 12}

Answer as indicated.

11. Determine the greatest fraction using mental math.

$$\frac{1790}{1791} \quad \frac{1789}{1790} \quad \frac{1789}{1792} \quad \frac{1788}{1789}$$

12. The sum of 2 positive consecutive integers is x. In terms of x, what fraction is the value of the lesser of these two integers?

13. If t orders of tea, each bought at c cents, are purchased per hour at a cafe that is open h hours a day, find the amount of money in dollars paid in 1 day for the tea.

14. A pig weighs 3/8 of a horse's weight. A cow weighs 6/5 of the horse's weight. The pig weighs 75 pounds. Find the weight of the cow in pounds.

15. In City C, 7/10 of the farmers grow only corn; the rest grow only wheat. Three-fifths of the corn farmers and 2/3 of the wheat farmers own their own land. Of the farmers in City C, what fraction own their own land?

Functions 1

Complete the function charts.

1. $f(x) = -2x^2 + 2x - 1$

x	0	1	−1	2	−2	3	−3	4	−4	5	−5	10	−10
$f(x)$													

2. $f(x) = 3x^2 - x + 4$

x	0	1	−1	2	−2	3	−3	4	−4	5	−5	10	−10
$f(x)$													

3. $f(x) = -x^2 + 3x - 2$

x	0	1	−1	2	−2	3	−3	4	−4	5	−5	10	−10
$f(x)$													

4. $f(x) = 2x^2 - 4x + 5$

x	0	1	−1	2	−2	3	−3	4	−4	5	−5	10	−10
$f(x)$													

5. $f(x) = -3x^2 + x - 3$

x	0	1	−1	2	−2	3	−3	4	−4	5	−5	10	−10
$f(x)$													

6. $f(x) = 4x^2 - 3x + 2$

x	0	1	−1	2	−2	3	−3	4	−4	5	−5	10	−10
$f(x)$													

7. $f(x) = 5x^2 - 5x + 6$

x	0	1	−1	2	−2	3	−3	4	−4	5	−5	10	−10
$f(x)$													

Functions 2

Answer YES or NO as to whether the relation is a function. If NO, state 2 ordered pairs that violate the function definition.

1. $y = 5x$	12. $y^2 = x^2$
2. $\{ (2, 3), (3, 3), (4, 3), (5, 3) \}$	13.
3. $y = x^2$	14. $\{ (1, 2), (2, 1), (4, 3), (3, 4) \}$
4. $x = y^2$	15. $x = 2y^2 + 4y - 2$
5. $2x - 7y = 1$	16.
6. $y = 5x^2 + 3x - 2$	17. $y = -5$
7. $y = x^3$	18. $x = 9$
8. $\{ (8, 6), (4, 2), (8, 3), (5, 1) \}$	19. $x = -4y + 12$
9. $x = 3y^2 - 1$	20. $x^2 + y^2 = 25$
10. $x = y^3$	21. $\{ (9, 0), (3, 2), (9, 2), (2, 0) \}$
11. $\{ (x, y) \mid x$ is a natural number, y is half $x\}$	22. $\{ (x, y) \mid x$ is an integer, y is twice x squared$\}$

Functions 3

Evaluate each function based on the given definitions.

ABS(x) = x if x > 0, –x if x < 0, and 0 if x = 0.
SGN(x) = 1 if x > 0, –1 if x < 0, and 0 if x = 0.
INT(x) = the greatest integer contained in x.
TRUNC(x) = x with its fractional or decimal part truncated.
SIGMA(x) = the number of divisors of x where x is a natural number.
SQR(x) = x squared.
SQRT(x) = the square root of x if x ≥ 0, undefined if x < 0.

1. ABS(–90)	14. TRUNC (7.75)	27. SQRT(0)
2. SQR(25)	15. SIGMA(8)	28. SQR(0)
3. TRUNC(8.97)	16. SQRT(–8)	29. ABS(0)
4. SIGMA(10)	17. INT(–9.2)	30. SIGMA(0)
5. SIGMA(1)	18. SGN(–32)	31. INT(0)
6. TRUNC(–53.5)	19. SQR(–11)	32. SGN(0)
7. INT(–53.5)	20. SQR(1.2)	33. TRUNC(–1.32)
8. SGN(987)	21. INT(–13)	34. ABS(–431)
9. SGN(3 – 17)	22. SIGMA(12)	35. SQR(0.5)
10. SIGMA(25)	23. TRUNC(–76.25)	36. TRUNC(9.97)
11. SQRT(81)	24. SGN(–3^5)	37. INT(–6.91)
12. ABS(5 – 9)	25. ABS(–5^3)	38. SGN(5 ÷ 3)
13. INT(21.5)	26. SQR(–6)	39. SQRT(1,000,000)

Functions 4

Write the linear function equation, and complete the function chart.

1. $f(x) =$

x	−2	−1	0	1	2	3	4
y			−5	−1			11

2. $f(x) =$

x	−2	−1	0	1	2	3	4
y	−4		2			11	

3. $f(x) =$

x	−2	−1	0	1	2	3	4
y			−4		8		20

4. $f(x) =$

x	−2	−1	0	1	2	3	4
y	−1		3		7		

5. $f(x) =$

x	−2	−1	0	1	2	3	4
y	−11		−1			14	

6. $f(x) =$

x	−2	−1	0	1	2	3	4
y			−8		6		20

7. $f(x) =$

x	−2	−1	0	1	2	3	4
y	15		7		−1		

8. $f(x) =$

x	−2	−1	0	1	2	3	4
y		−3	−6			−15	

9. $f(x) =$

x	−2	−1	0	1	2	3	4
y			9		5	3	

10. $f(x) =$

x	−2	−1	0	1	2	3	4
y	11			1	−4		

11. $f(x) =$

x	−2	−1	0	1	2	3	4
y		−11	−3				29

12. $f(x) =$

x	−2	−1	0	1	2	3	4
y		−4	5	14			

Functions 5

Answer YES or NO as to whether the graph depicts a function.

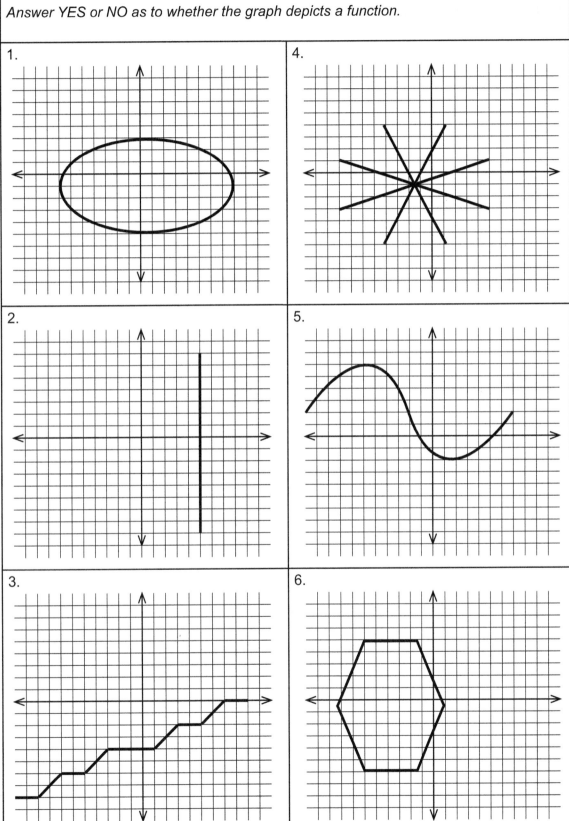

1.

2.

3.

4.

5.

6.

Functions 6

Evaluate using linear interpolation.

1. $f(10) = 500$ $f(13) = 260$ Find $f(12.5)$	8. $f(14) = 48.6$ $f(16) = 22.2$ Find $f(15.5)$
2. $f(6) = 3.087$ $f(8) = 1.871$ Find $f(6.5)$	9. $f(6) = 39.043$ $f(12) = 2.677$ Find $f(7)$
3. $f(2) = 17.09$ $f(6) = 33.97$ Find $f(4.5)$	10. $f(22) = 22.74$ $f(27) = 67.89$ Find $f(25)$
4. $f(20) = 769$ $f(30) = 354$ Find $f(23)$	11. $f(6) = 5306$ $f(7) = 8541$ Find $f(6.7)$
5. $f(2) = 32.4$ $f(3) = 53.5$ Find $f(2.9)$	12. $f(5.1) = 9465$ $f(5.4) = 3762$ Find $f(5.2)$
6. $f(31) = 3084$ $f(36) = 2269$ Find $f(33)$	13. $f(7) = 36.22$ $f(9) = 22.08$ Find $f(8)$
7. $f(6) = 72.9$ $f(8) = 24.1$ Find $f(7.25)$	14. $f(5) = 6850$ $f(7) = 4390$ Find $f(6.7)$

Functions 7

Evaluate.	*Find k, the unknown constant term, given one value of the quadratic function.*

1. $f(x) = 3x + 2$
    $g(x) = 2x - 1$
    Find $g(f(4)) + f(g(2))$.
	12. $f(x) = 3x^2 + 2x + k$
	    $f(2) = 6$

2. $f(x) = 4x - 5$
    $g(x) = 5x + 6$
    Find $g(f(2)) + f(g(1))$.
	13. $f(x) = 5x^2 + 3x - k$
	    $f(3) = 50$

3. $f(x) = 3x^2 - 1$
    $g(x) = 2x + 3$
    Find $g(f(2)) - f(g(-2))$.
	14. $f(x) = -x^2 + 4x - k$
	    $f(4) = 8$

4. $f(x) = x^x$
    $g(x) = 3x + 1$
    Find $g(f(2)) + f(g(1))$.
	15. $f(x) = 2x^2 + 3x - k$
	    $f(-2) = 10$

5. $f(x) = 2x^2 + 3$
    $g(x) = 5x + 6$
    Find $g(f(3)) - f(g(-3))$.
	16. $f(x) = -4x^2 - 3x - k$
	    $f(-2) = 5$

6. $f(x) = (x - 3) \div 2$
    $g(x) = 4x + 5$
    Find $g(f(7)) - f(g(1))$.
	17. $f(x) = 6x^2 + 5x + k$
	    $f(-5) = 100$

7. $f(x) = 3x^2 + 2x - 3$
    $g(x) = -3x + 8$
    Find $g(f(-2)) + f(g(1))$.
	18. $f(x) = 4x^2 + 2x - k$
	    $f(-3) = 14$

8. $f(x) = 2x - 3$
    $g(x) = 4x + 3$
    Find $g(f(2)) + f(g(2))$.
	19. $f(x) = -x^2 + 7x + k$
	    $f(-4) = 4$

9. $f(x) = (3x + 1) \div 2$
    $g(x) = (2x - 1) \div 3$
    Find $g(f(3)) + f(g(2))$.
	20. $f(x) = -4x^2 - 2x + k$
	    $f(-1) = 7$

10. $f(x) = x^{-x}$
    $g(x) = 8x + 6$
    Find $g(f(2)) \div f(g(-1))$.
	21. $f(x) = 5x^2 + 6x - k$
	    $f(1) = 4$

11. $f(x) = (-x)^3$
    $g(x) = -2x + 6$
    Find $g(f(3)) + f(g(1))$.
	22. $f(x) = -3x^2 - 5x + k$
	    $f(-2) = 6$

Functions 8

Evaluate the recursive function, given the domain is the set of natural numbers.

1. $f(n) = 3f(n-1) - n$
   $f(1) = 4$
   Find $f(4)$.

2. $f(mn) = f(m) + f(n)$
   $f(8) = 20$
   Find $f(64)$.

3. $f(n+2) = 5f(n) - 5$
   $f(1) = 5$
   Find $f(7)$.

4. $f(mn) = f(m) - f(n)$
   Find $f(81)$.

5. $f(n) = (f(n-1))^2 - 2$
   $f(3) = 10$
   Find $f(0)$.

6. $f(n) = (f(n+1))^2$
   $f(4) = 10$
   Find $f(1)$.

7. $f(n) = 2f(n+1) + 3$
   $f(10) = 30$
   Find $f(7)$.

8. $f(n) = nf(n+1)$
   $f(4) = 5$
   Find $f(1)$.

9. $f(n) = f(n+2) + 1$
   $f(7) = 10$
   Find $f(1)$.

10. $f(n) = 4f(n-1) + 2n$
    $f(1) = 4$
    Find $f(4)$.

11. $f(n) = 4f(n-2)$
    $f(0) = 0, f(1) = 2$
    Find $f(5)$.

12. $f(n) = 1 + f(3n-1)$ if n odd and n>1
    $f(n) = 2n$ if n even
    $f(1) = 1$
    Find $f(3)$.

13. $f(n) = 2 + f(5n-1)$ if n odd and n>1
    $f(n) = n \div 2$ if n even
    $f(1) = 1$
    Find $f(5)$.

14. $f(n) = f(n+1) - 2$ if n odd and n>1
    $f(n) = 3n$ if n even
    $f(1) = 5$
    Find $f(7)$.

15. $f(n) = 2f(n-1) - f(n-2)$
    $f(1) = 4, f(2) = 9$
    Find $f(5)$.

16. $f(n) = -3f(n-1) - 3f(n-2)$
    $f(1) = 2, f(2) = 5$
    Find $f(5)$.

Geometric Structures 1

Name with proper notation, using the diagrams.	*Name with proper notation, using the diagram. (Answers may vary.)*
1.　Name 6 different rays. 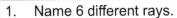	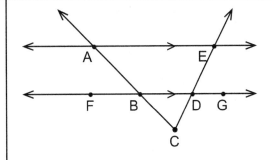
	5.　3 noncollinear points
	6.　4 collinear points
2.　Name 6 different angles. 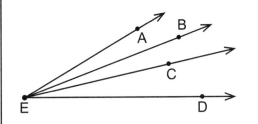	7.　1 scalene trapezoid
	8.　2 parallel rays
	9.　2 parallel line segments
	10.　4 line segments with endpoint C
	11.　2 vertical angles
3.　Name 9 different line segments. 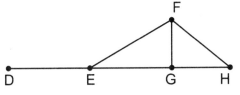	12.　2 supplementary angles
	13.　2 alternate interior angles
	14.　1 obtuse angle
	15.　1 acute angle
4.　Name 4 different line segments that all intersect with each other. 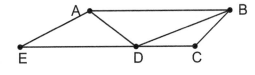	16.　a ray containing B, not as an endpoint
	17.　2 rays intersecting in infinitely many points
	18.　2 similar triangles
	19.　2 congruent line segments
	20.　3 rays with endpoint B

Geometric Structures 2

Answer for 4 distinct, coplanar points A, B, C, and D, only pairwise collinear.

1. Two lines always intersect in one point.

12. How many lines contain A and B?

2. Two different lines may never intersect in more than one point.

13. How many lines contain A, B, and C?

3. Two non-parallel lines can have an empty intersection.

14. How many lines contain A, B, C, and D?

4. Two line segments may never intersect in more than one point.

15. How many planes contain A?

5. Two non-parallel line segments in a plane can have an empty intersection.

16. How many planes contain A and D?

6. Two planes may intersect in one point.

17. How many planes contain A, B, and D?

7. If two planes do not intersect, they are parallel.

18. How many planes contain A, B, C, and D?

8. The intersection of a line and a plane is a point.

19. If two lines intersect at D, at how many other points do they intersect?

9. Three parallel lines must be coplanar.

20. The line through A and B and the line through C and D intersect in how many points?

10. If two points of a ray are contained in a line, the entire ray is contained in the line.

21. Angle BAD and angle BCD intersect in how many points?

11. If two points of a line segment are contained in a ray, the entire segment is contained in the ray.

22. The intersection of 2 planes contains B and how many other points?

Graphs 1

Graph on the number line.

1. x = 5

    ← —————————————→

2. x > 7

    ← —————————————→

3. x ≤ 4

    ← —————————————→

4. −4 < x < 9

    ← —————————————→

5. 0 ≤ x ≤ 3

    ← —————————————→

6. x > 4

    ← —————————————→

7. x = 0

    ← —————————————→

8. x ≥ 5.5

    ← —————————————→

9. −1 < x < 1

    ← —————————————→

10. x = 2.5

    ← —————————————→

11. −2 ≤ x < 3

    ← —————————————→

12. x < 8

    ← —————————————→

13. x ≥ 3

    ← —————————————→

14. x = − 4

    ← —————————————→

15. −3 < x ≤ 5

    ← —————————————→

16. −5 ≤ x ≤ 3

    ← —————————————→

17. −1 < x < 5

    ← —————————————→

18. x ≤ 7

    ← —————————————→

19. x > −4

    ← —————————————→

20. −4.5 < x < 4.5

    ← —————————————→

21. −2 ≤ x < 6

    ← —————————————→

22. x < −2

    ← —————————————→

Graphs 2

Without graphing, identify as a point, two points, line, line segment, open line segment, half-open line segment, ray, open ray, two rays, or CBD if graphed on the number line.

1. $x \geq 8$	17. $x = -7$ OR $x = -8$
2. $x = -9$	18. $0 \leq x \leq 99$
3. $-12 \leq x \leq 15$	19. $x \geq 34$
4. $x = 4.3$ OR $x = -5.2$	20. $x \geq 11$ OR $x \geq 13$
5. $-10 < x < 11$	21. $x = 25$ OR $x = -32$
6. $2x = 10$	22. $-7 < x < 14$
7. $x + 2 = 5 + x - 3$	23. $35 \geq x \geq 22$
8. $x = 2$ AND $x = 8$	24. $x = -109.6$
9. $x > 19$	25. $x < 5$ OR $x > -1$
10. $x \leq -13$	26. $x > -23$
11. $x < 2$ OR $x > 0$	27. $x \geq 0$ OR $x \leq -6$
12. $x < 12$	28. $3x + 1 = 8$
13. $x + 1 = 6$	29. $0.9 \leq x \leq 1.3$
14. $-4 < x \leq 24$	30. $x \leq 59$ OR $x \leq 46$
15. $x > 1$ AND $x < 0$	31. $x < 0$ OR $x \geq 0$
16. $x \leq -5$ OR $x \geq 17$	32. $-11 \leq x < -1$

Graphs 3

Graph on the number line.

1. x = 4 OR x = –3

←———————————————→

12. x = 3 AND x = 4

←———————————————→

2. x > 6 OR x < 2

←———————————————→

13. x ≥ 3 AND x < 0

←———————————————→

3. x ≤ 1 OR x > 4

←———————————————→

14. x ≤ 0 OR x > 3

←———————————————→

4. x = 3 OR x = –4 OR x = 6

←———————————————→

15. –3 ≤ x AND x ≤ 6

←———————————————→

5. –5 ≤ x AND x ≤ 4

←———————————————→

16. x = 5 OR x = –2

←———————————————→

6. x < 0 OR x > 2

←———————————————→

17. –1 < x OR –3 < x

←———————————————→

7. x > 4 OR x ≥ 3

←———————————————→

18. x < –2 OR x = 8

←———————————————→

8. x = 3 OR x > 7

←———————————————→

19. x = –4 OR x = 0 OR x = 4

←———————————————→

9. x < –3 OR x > 5

←———————————————→

20. –1 < x AND x < 5

←———————————————→

10. x = 1.75 AND x = 2.5

←———————————————→

21. x > 6 OR x > 4

←———————————————→

11. –3 ≤ x AND x < 3

←———————————————→

22. x = –2 OR x > 0

←———————————————→

Graphs 4

Compute the slope of the line determined by the two points.

1. (0, 0)　　(6, 4)	12. (3, 3)　　(2, −2)	
2. (−1, −6)　(3, 2)	13. (1, 4)　　(6, −1)	
3. (5, 3)　　(4, 1)	14. (6, 7)　　(−3, −2)	
4. (−2, −1)　(0, 4)	15. (−3, −4)　(5, −8)	
5. (−7, 6)　　(6, 6)	16. (−5, 2)　(3, −3)	
6. (−3, 0)　　(2, 3)	17. (8, 0)　　(1, −7)	
7. (−5, −6)　(2, −4)	18. (−2, −2)　(8, 3)	
8. (0, −4)　　(5, 1)	19. (−3, 2)　(−2, −5)	
9. (0, 0)　　(−6, 4)	20. (8, 4)　　(2, 0)	
10. (0, −8)　　(11, 3)	21. (−5, 7)　(−5, −8)	
11. (4, 3)　　(−4, 5)	22. (3, 3)　　(5, 5)	

Graphs 5

Complete the chart. Graph by plotting 3 points. Draw the line using a straightedge, and cross both axes.

1. y = 4x − 1

x	y
−3	
−2	
−1	
	−1
1	
	7
3	

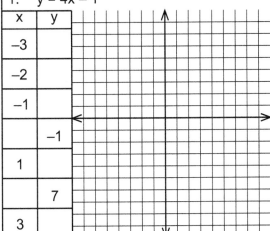

4. y = −2x + 1

x	y
−3	
	5
−1	
0	
1	
2	
3	

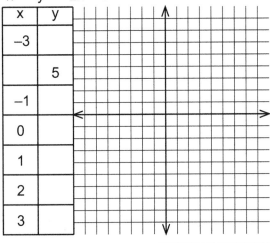

2. y = 2x + 3

x	y
−2	
−1	
0	
	5
	7
3	
4	

5. y = −3x − 2

x	y
−2	
−1	
0	
1	
	−8
3	
4	

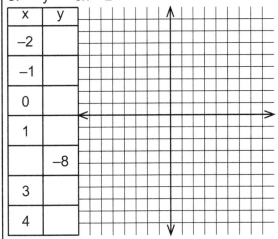

3. y = 3x − 3

x	y
−2	
−1	
0	
	0
	3
3	
4	

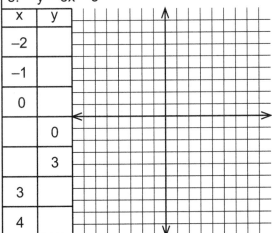

6. y = −x + 4

x	y
−3	
	6
−1	
0	
1	
2	
5	

Graphs 6

Complete the chart for the linear equation.

	EQUATION	SLOPE	X-INT	Y-INT		EQUATION	SLOPE	X-INT	Y-INT
1.	$y = 2x - 2$				17.	$y = 6x + 5$			
2.	$y = 3x$				18.	$3x + 7y = 21$			
3.	$x = -5$				19.	$8y = 4x - 3$			
4.	$y = 4x + 4$				20.	$5y = x + 3$			
5.	$2y = 14$				21.	$x = 3y + 7$			
6.	$3y = 2x - 6$				22.	$2x = y - 1$			
7.	$y = 7x - 3$				23.	$y = 9$			
8.	$y = 5x + 1$				24.	$y - x = 4$			
9.	$3y = 12x - 7$				25.	$x = 4y$			
10.	$3x + 4y = 12$				26.	$x + y = 10$			
11.	$y = 5x + 9$				27.	$2y = 6x - 1$			
12.	$2x - 2y = 7$				28.	$x + 5y = 4$			
13.	$2y = -9x - 2$				29.	$4x + y = -5$			
14.	$x - y = 6$				30.	$y = -7x$			
15.	$2x + 5y = 20$				31.	$x = 6$			
16.	$5x + 3y = 30$				32.	$2y = 8x - 3$			

Graphs 7

Write the equation of the line in slope-intercept form.

1.   y =

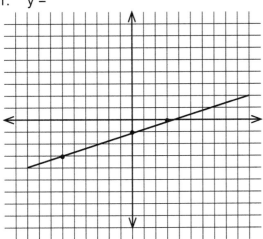

4.   y =

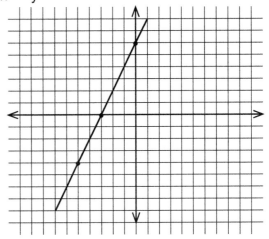

2.   y =

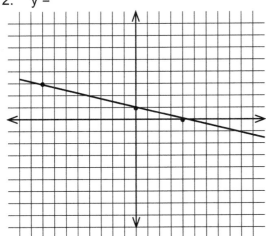

5.   y =

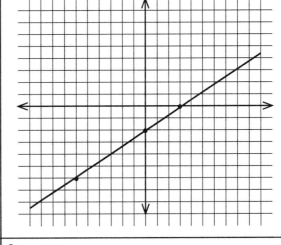

3.   y =

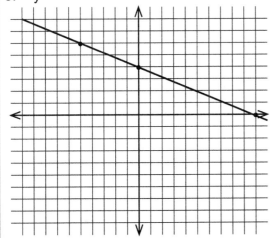

6.   y =

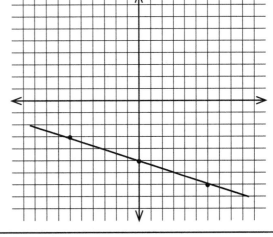

MAVA Math: Enhanced Skills   Copyright © 2015 Marla Weiss

Graphs 8

Graph by plotting points.

1. y = |x|

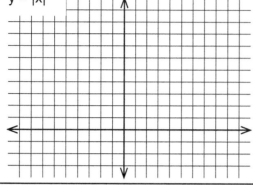

2. y = |x| + 1

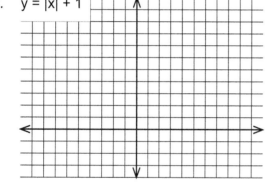

3. y = |x| − 2

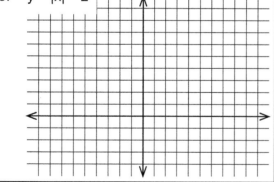

4. y = − |x|

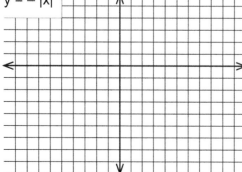

5. y = |3x|

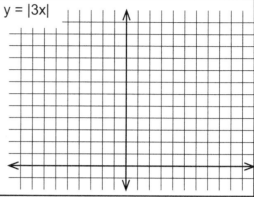

6. x = |y|

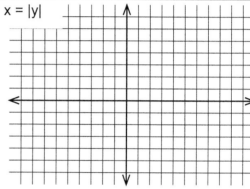

7. x = − |y|

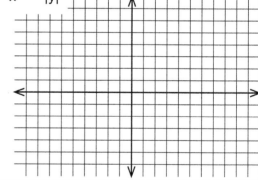

8. y = |2x + 2|

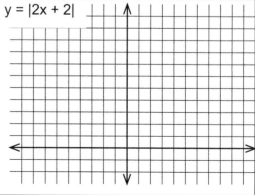

Graphs 9

Graph each line with a straightedge. Use the slope and y-intercept.

1. $y = 3x + 5$

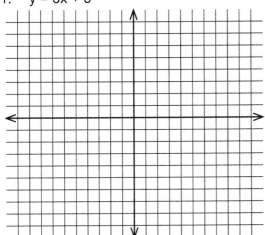

4. $y = \dfrac{-2}{3}x + 6$

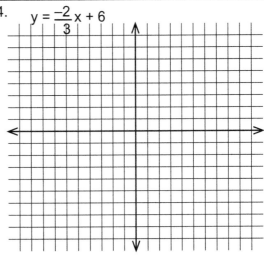

2. $y = -4x - 4$

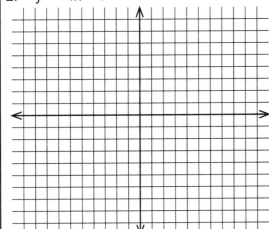

5. $y = \dfrac{3}{4}x - 3$

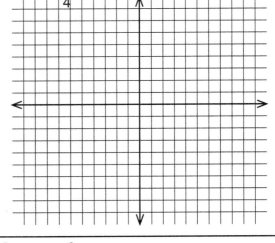

3. $y = 2x + 3$

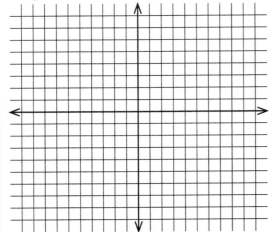

6. $y = \dfrac{-2}{5}x - 2$

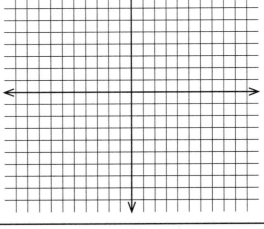

Graphs 10

Graph the circle. Identify the center (C). Label four lattice points on the circle.

1. $x^2 + y^2 = 16$

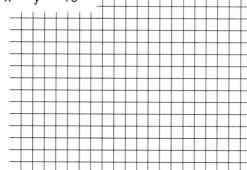

5. $(x - 5)^2 + y^2 = 36$

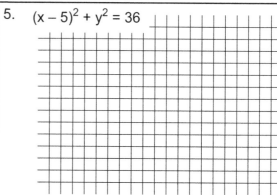

2. $x^2 + y^2 = 36$

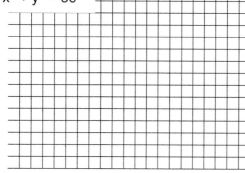

6. $x^2 + (y - 2)^2 = 9$

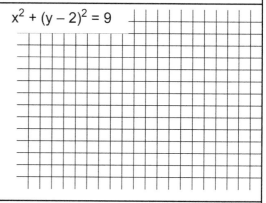

3. $x^2 + (y - 1)^2 = 49$

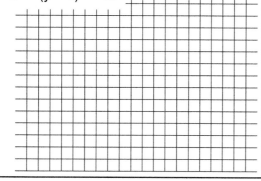

7. $(x - 2)^2 + (y - 3)^2 = 16$

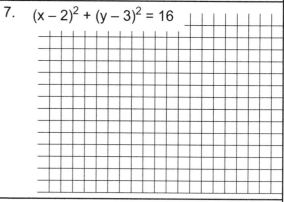

4. $(x - 3)^2 + y^2 = 25$

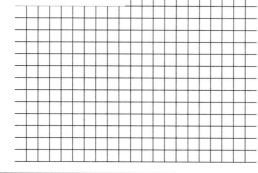

8. $(x - 1)^2 + (y - 4)^2 = 9$

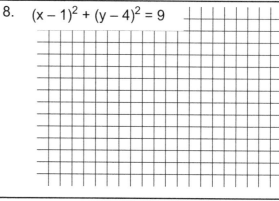

Graphs 11

Write a linear equation given two intercepts.	*Write a linear equation given slope and one intercept.*
1. (0, 3) and (6, 0)	9. slope = 5; y-intercept = (0, 4)
2. (0, 7) and (2, 0)	10. slope = 4; x-intercept = (2, 0)
3. (0, −4) and (5, 0)	11. slope = 2/3; y-intercept = (0, −5)
4. (0, 2) and (−8, 0)	12. slope = 2; x-intercept = (−3, 0)
5. (0, 5) and (6, 0)	13. slope = −3/5; y-intercept = (0, 1)
6. (0, −2) and (7, 0)	14. slope = −3; x-intercept = (1, 0)
7. (0, 9) and (9, 0)	15. slope = 3/4; y-intercept = (0, −2)
8. (0, 1) and (−7, 0)	16. slope = 6; x-intercept = (0, 0)

Graphs 12

Write a linear equation given two points.	Write a linear equation given slope and one point.
1. (3, 6) and (4, 7)	9. slope = 3; point = (1, 5)
2. (4, 5) and (6, 9)	10. slope = –4; point = (2, –7)
3. (3, –4) and (5, –6)	11. slope = 2/3; point = (6, 9)
4. (2, 4) and (4, 7)	12. slope = 3/4; point = (8, –3)
5. (3, 7) and (6, 9)	13. slope = –1/5; point = (–10, 4)
6. (6, –3) and (8, –8)	14. slope = 5/6; point = (12, 7)
7. (6, 3) and (9, 7)	15. slope = –3/7; point = (–14, 9)
8. (1, –10) and (–1, –2)	16. slope = 7/8; point = (8, 6)

Greatest Common Factor 1

Find the GCF mentally.

1. 45 and 81	17. 11, 13, and 31	33. 36 and 70	49. 49 and 50
2. 12 and 72	18. 550 and 600	34. 8, 18, and 28	50. 21 and 91
3. 17 and 64	19. 230 and 590	35. 6, 18, and 42	51. 82 and 86
4. 28 and 70	20. 24, 30, and 39	36. 120 and 330	52. 624 and 625
5. 18 and 32	21. 22 and 451	37. 50, 90, and 360	53. 21 and 1001
6. 18, 45, and 90	22. 100 and 150	38. 8, 27, and 64	54. 91 and 1001
7. 5, 7, and 20	23. 56 and 140	39. 121 and 385	55. 84 and 88
8. 6, 9, and 15	24. 12, 15, and 16	40. 18, 30, and 66	56. 60 and 75
9. 24 and 42	25. 56 and 112	41. 31 and 62	57. 220 and 242
10. 26 and 56	26. 38, 57, and 190	42. 8, 10, and 12	58. 150 and 360
11. 34 and 51	27. 22, 33, and 55	43. 32 and 160	59. 200 and 260
12. 36, 54, and 90	28. 21 and 56	44. 5, 6, and 21	60. 33 and 363
13. 38 and 56	29. 5, 10, and 15	45. 4, 11, and 16	61. 120 and 330
14. 72 and 90	30. 27 and 32	46. 5, 6, and 10	62. 27 and 51
15. 6, 7, and 55	31. 12, 30, and 42	47. 18 and 45	63. 198 and 374
16. 30, 42, and 48	32. 9, 15, and 33	48. 44, 66, and 88	64. 35, 42, and 77

Greatest Common Factor 2

Find the greatest common factor.

1. $GCF(a^3b^4c^5 , a^2b^6c^3)$

2. $GCF(16x^4y^5z^3 , 48x^3y^6z^3)$

3. $GCF(5w^7x^4y^5z^3 , 7w^2x^5y^4z)$

4. $GCF(5a^3s^4 , 6a^4s^3 , 12a^2s^3)$

5. $GCF(35a^2b^2c^2 , 42a^3b^5)$

6. $GCF(6f^7g^3h^6 , 10f^4g^5h^8)$

7. $GCF(36d^5e^2f^6 , 144d^3e^6f^8)$

8. $GCF(5x^4y^5z^6 , 51x^3y^2z^8)$

9. $GCF(12rst , 16r^2s^5t^4 , 8r^3s^3t^6)$

10. $GCF(6b^3d^5g^6 , 9b^4d^5h^2)$

11. $GCF(4d^2e^3 , 5d^4e^4 , 6d^6e^5)$

12. $GCF(15x^9y^6z^5 , 70x^5y^6z^9)$

13. $GCF(11a^4b^4c^3 , 44a^9b^3c^6)$

14. $GCF(5g^7s^6 , 10g^6s^7 , 15g^8s^3)$

15. $GCF(36p^4q^2r^8 , 90p^5q^2r^3)$

16. $GCF(14a^5bc^7 , 63a^3b^9)$

17. $GCF(14f^7g^6h^4 , 16f^3g^4h^2)$

18. $GCF(16r^4st^2 , 28r^6s^3t^5 , 4r^2s^3t)$

19. $GCF(15x^2y^3z^4 , 20w^2y^4z^3)$

20. $GCF(14p^9q^2r^4 , 70p^5q^5r^6)$

21. $GCF(6m^3n^6p^7 , 15m^2n^4p^8)$

22. $GCF(6e^3f^5g^5 , 8e^6f^5g^7)$

Greatest Common Factor 3

Find the greatest common factor by prime factorization.

1.	130	286	9.	780	910	
2.	405	495	10.	9000	2772	
3.	819	1260	11.	3850	2205	
4.	1078	2450	12.	80	96	144
5.	4200	5000	13.	78	102	726
6.	341	1001	14.	24	54	72
7.	396	2070	15.	132	165	297
8.	2700	1080	16.	84	168	210

Greatest Common Factor 4

Given one of two numbers m and their GCF and LCM, find the second number n.	*Evaluate as indicated.*
1. m = 20 GCF(m, n) = 10 LCM(m, n) = 60	9. LCM(GCF(45, 63), GCF(36, 42))
2. m = 60 GCF(m, n) = 30 LCM(m, n) = 4620	10. LCM(GCF(28, 42), GCF(45, 60))
3. m = 154 GCF(m, n) = 7 LCM(m, n) = 8008	11. GCF(LCM(30, 75), LCM(70, 45))
4. m = 48 GCF(m, n) = 12 LCM(m, n) = 432	12. LCM(GCF(120, 168), GCF(28, 92))
5. m = 288 GCF(m, n) = 18 LCM(m, n) = 1440	13. GCF(LCM(91, 49), LCM(222, 66))
6. m = 72 GCF(m, n) = 24 LCM(m, n) = 3960	14. LCM(GCF(147, 210), GCF(56, 64))
7. m = 210 GCF(m, n) =15 LCM(m, n) = 7350	15. GCF(LCM(30, 99), LCM(55, 60))
8. m = 198 GCF(m, n) = 18 LCM(m, n) = 4158	16. LCM(GCF(225, 825), GCF(70, 84))

Historical Math 1

Use the Sieve of Eratosthenes to determine all prime numbers listed. Circle a prime; then cross off all of its multiples. Proceed to the next prime, and repeat.

1	2	3	4	5	6	7	8	9	10	11	12
13	14	15	16	17	18	19	20	21	22	23	24
25	26	27	28	29	30	31	32	33	34	35	36
37	38	39	40	41	42	43	44	45	46	47	48
49	50	51	52	53	54	55	56	57	58	59	60
61	62	63	64	65	66	67	68	69	70	71	72
73	74	75	76	77	78	79	80	81	82	83	84
85	86	87	88	89	90	91	92	93	94	95	96
97	98	99	100	101	102	103	104	105	106	107	108
109	110	111	112	113	114	115	116	117	118	119	120
121	122	123	124	125	126	127	128	129	130	131	132
133	134	135	136	137	138	139	140	141	142	143	144
145	146	147	148	149	150	151	152	153	154	155	156
157	158	159	160	161	162	163	164	165	166	167	168
169	170	171	172	173	174	175	176	177	178	179	180
181	182	183	184	185	186	187	188	189	190	191	192
193	194	195	196	197	198	199	200	201	202	203	204
205	206	207	208	209	210	211	212	213	214	215	216
217	218	219	220	221	222	223	224	225	226	227	228
229	230	231	232	233	234	235	236	237	238	239	240
241	242	243	244	245	246	247	248	249	250	251	252
253	254	255	256	257	258	259	260	261	262	263	264

Historical Math 2

Answer YES or NO as to whether the figure is an Euler graph (may be traced drawing each line exactly once without lifting the pencil). Mark the start and stop.

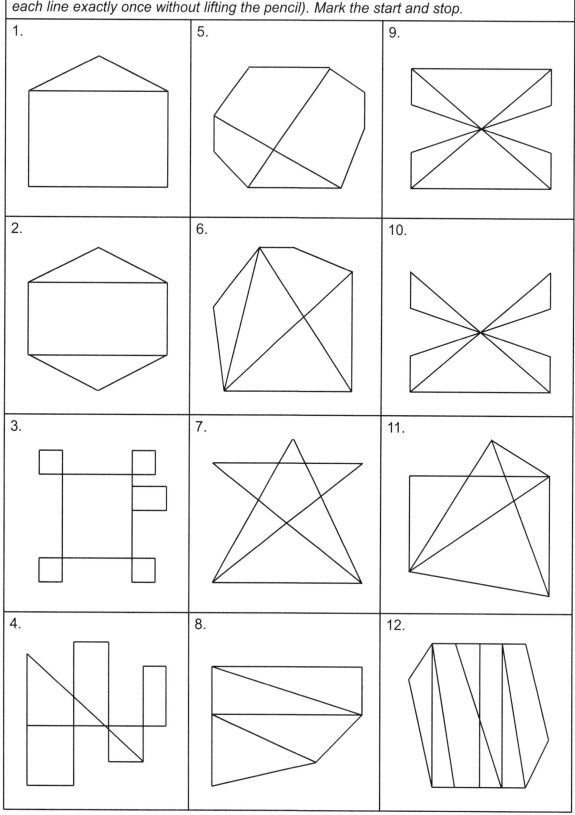

Historical Math 3

Verify Goldbach's conjecture (through 130) that every even integer greater than 2 is the sum of two primes. Answers may vary.

1. 4	17. 36	33. 68	49. 100
2. 6	18. 38	34. 70	50. 102
3. 8	19. 40	35. 72	51. 104
4. 10	20. 42	36. 74	52. 106
5. 12	21. 44	37. 76	53. 108
6. 14	22. 46	38. 78	54. 110
7. 16	23. 48	39. 80	55. 112
8. 18	24. 50	40. 82	56. 114
9. 20	25. 52	41. 84	57. 116
10. 22	26. 54	42. 86	58. 118
11. 24	27. 56	43. 88	59. 120
12. 26	28. 58	44. 90	60. 122
13. 28	29. 60	45. 92	61. 124
14. 30	30. 62	46. 94	62. 126
15. 32	31. 64	47. 96	63. 128
16. 34	32. 66	48. 98	64. 130

Historical Math 4

Verify the first 4 pairs of amicable numbers by showing the sum of the proper divisors of each equals the other.

Verify the first 4 perfect numbers by summing proper divisors.

1. 220	284	5. 6
2. 1184	1210	6. 28
3. 2620	2924	7. 496
4. 5020	5564	8. 8128

Inequalities 1

Solve.

1. $5x < 35$	12. $9 - x \geq -11$	23. $3 - 10x \geq 36 + x$
2. $3x \leq 33$	13. $10 - 3x < 40$	24. $-7x \leq x + 88$
3. $-7x > -63$	14. $3x - 1 \leq -40$	25. $9x - 2 \geq -12 - x$
4. $13x \geq -52$	15. $11x + 4 \leq -28 + 3x$	26. $100 > 37 - 7x$
5. $2x + 3 < -55$	16. $11 - x > 44$	27. $5x + 5 < x - 43$
6. $-x - 2 \leq 24$	17. $8x + 5 \leq -27$	28. $3x - 8 \geq 32 - 5x$
7. $4x + 6 > 7x$	18. $-6x + 1 \geq -71$	29. $12 - 5x < -48 + x$
8. $6x + 11 < 3x - 1$	19. $4x < -x + 55$	30. $8 - 7x > -48$
9. $-2x + 7 \leq 39 - x$	20. $9x - 4 \leq -28 + x$	31. $3 - 2x \leq 45 + x$
10. $8x + 5 < -61$	21. $4 - x \geq x - 22$	32. $9x + 5 > 65$
11. $-10x - 8 \leq -73$	22. $3x - 7 \leq 35 + 9x$	33. $-5x - 4 < 64 + 3x$

Inequalities 2

Solve.

1. $\dfrac{1}{6}x \geq -12$	9. $\dfrac{3}{4}(8x - 4) \geq 21$	17. $\dfrac{2}{3}(5x - 1) \geq 11$
2. $\dfrac{2}{5}x + 1 < 31$	10. $\dfrac{2}{7}(14x + 21) < -5$	18. $\dfrac{4}{5}(8x + 3) < -4$
3. $\dfrac{3}{8}x - 6 \leq -30$	11. $\dfrac{-3}{14}(42x - 14) > 11$	19. $\dfrac{-3}{7}(5x - 8) > 12$
4. $2 - \dfrac{5}{7}x > -43$	12. $\dfrac{11}{15}(45 + 30x) \leq 44$	20. $\dfrac{5}{8}(9 - 5x) \leq 15$
5. $10 - \dfrac{5}{9}x < 25$	13. $\dfrac{-4}{21}(63x) \geq 22 - x$	21. $\dfrac{-3}{4}(10x) \geq -14 - 3x$
6. $\dfrac{1}{7}x - 2 \geq 9$	14. $\dfrac{5}{9}(18x - 27) \leq 5x$	22. $\dfrac{5}{6}(9x - 18) \leq 15x$
7. $\dfrac{2}{3}x + 5 < -13$	15. $\dfrac{3}{5}(45x + 20) \geq 3x$	23. $\dfrac{1}{2}(17x + 18) \geq 4x$
8. $\dfrac{11}{20}x - 2 \geq 86$	16. $\dfrac{-5}{13}(13 + 26x) > -25$	24. $\dfrac{-4}{9}(6 + 20x) > -16$

Integers 1

Add or subtract as indicated. Use mental math.

1. 15 – 30	17. 11 – 20	33. 50 – 90	49. –34 + 87
2. –20 + (–11)	18. –35 + –55	34. –15 + –46	50. –45 + –54
3. –45 + 45	19. 29 – 50	35. 25 – 66	51. 20 – 72
4. 13 – 18	20. 0 – 99	36. –23 + –67	52. 33 + –79
5. –30 – (–12)	21. –23 – (–54)	37. 20 – 30	53. 33 – 79
6. –16 – 16	22. 53 – 70	38. –53 + –46	54. –33 + –46
7. 14 – 21	23. 88 + –77	39. –12 + –79	55. 46 + –79
8. –89 + 13	24. –76 + 78	40. 11 – 16	56. 79 – 46
9. 89 – 13	25. –17 + –18	41. –19 – 15	57. –47 – 27
10. –13 + 89	26. 52 – 98	42. 40 – 48	58. 30 – 88
11. 13 – 89	27. 11 – 13	43. –18 + –19	59. –26 + –36
12. 41 – (–47)	28. 19 – 16	44. –14 + –53	60. –28 + –28
13. –34 – 36	29. 16 – 19	45. 25 – 35	61. 37 – 70
14. 84 + (–52)	30. –19 + –16	46. –55 + 59	62. –19 + 89
15. 45 – 50	31. –16 + 19	47. 18 – 36	63. 30 – 55
16. –51 + 51	32. –19 + 16	48. –15 + 75	64. 42 + –90

Integers 2

Multiply or divide as indicated. Use mental math.

1. 19 x (– 2)	17. (–101)(–9)	33. –121 ÷ 11
2. –17 x (–4)	18. (–20)(–30)(–40)	34. 64 ÷ (–16)
3. –20 x 60	19. –4907 ÷ 7	35. (–11)(–16)
4. –51 ÷ 3	20. –21 x 5	36. –45 ÷ (–3)
5. (–4)(–7)(–5)	21. (–35)(–3)	37. (–99) ÷ (–9)
6. (–16)(–3)	22. (–5)(6)(–25)	38. –169 ÷ 13
7. (–12)(–7)(–5)	23. –17 x 3	39. –810 ÷ (–9)
8. (–8)(6)(–5)	24. –600 ÷ 15	40. –743 x –10
9. –93 ÷ 3	25. –560 ÷ (–70)	41. –369 ÷ 3
10. –45 x 11	26. –6 x 13	42. –816 ÷ (–4)
11. (–880) ÷ (–11)	27. –824 ÷ (–4)	43. –11 x –27
12. –9036 ÷ 9	28. (–53)(–11)	44. (–100)(–10)
13. –150 ÷ (–30)	29. 1000 ÷ (–5)	45. (–12)(–12)
14. 102 ÷ (–3)	30. (–15)(–7)	46. 444 ÷ (–4)
15. –105 ÷ 5	31. 25 x (–8)	47. –36 ÷ (–2)
16. –(33 x 5)	32. –864 ÷ 8	48. 43 x (–11)

Integers 3

Find two consecutive positive integers with the given description.	*Find consecutive integers with the given description.*
1. sum is 37	20. 2 consecutive even, sum is 54
2. half their product is 45	21. 5 consecutive odd, sum is 15
3. triple their greater quotient is 4	22. 5 consecutive even, sum is 0
4. twice their sum is 22	23. 3 consecutive odd, sum is 69
5. sum is 53	24. 4 consecutive even, sum is 36
6. half their product is 66	25. 6 consecutive, sum is −9
7. average is 31.5	26. 6 consecutive odd, sum is 0
8. triple their sum is 105	27. 9 consecutive, sum is −9
9. product is 2550	28. 6 consecutive even, sum is −6
10. double their lesser quotient is 1	29. 8 consecutive, sum is 12
11. half their sum is 15.5	30. 8 consecutive odd, sum is −16
12. average is 89.5	31. 5 consecutive odd, sum is 75
13. product is 650	32. 7 consecutive odd, sum is 21
14. sum is 151	33. 4 consecutive even, sum is 140
15. double their average is 27	34. 8 consecutive, sum is −4
16. product is 600	35. 5 consecutive odd, sum is 465
17. half their product is 55	36. 9 consecutive, sum is −18
18. product is 210	37. 3 consecutive odd, sum is 153
19. triple their sum is 135	38. 7 consecutive even, sum is −14

Integers 4

Answer as indicated.

1. Find 4 consecutive odd integers such that the sum of the two greatest subtracted from twice the sum of the two least is 76.	6. Find 5 consecutive multiples of 3 such that twice the sum of the 2nd and 4th is 30 more than the sum of the greatest and least.
2. Find 5 consecutive odd integers such that the 4th is the sum of the 1st and twice the 3rd.	7. Find 6 consecutive integers such that twice the sum of the 2 least minus the sum of the 2 greatest is 33.
3. Find 3 consecutive multiples of 4 such that triple the 3rd minus twice the 2nd is 4 less than twice the 1st.	8. Find 4 consecutive even integers such that the sum of the 3 greatest is 90 more than the sum of the 2 least.
4. Find 4 consecutive integers such that 4 times the 3rd decreased by triple the 2nd is 39.	9. Find 4 consecutive multiples of 5 such that the sum of the 3 least is 20 more than twice the 4th.
5. Find 3 consecutive even integers such that the sum of the least and greatest is 108.	10. Find two consecutive even integers such that twice the greater is 22 more than triple the lesser.

Interest 1

Complete the chart of annual simple interest by mental math.

	PRINCIPAL	RATE	INTEREST		PRINCIPAL	RATE	INTEREST
1.	$100	6%		20.	$1000	18%	
2.		7%	$14	21.		10%	$45
3.	$250	10%		22.	$350	20%	
4.		6%	$27	23.	$275		$22
5.		8%	$24	24.		5%	$450
6.	$300		$33	25.	$6000		$180
7.	$500		$75	26.	$5200		$104
8.	$400	17%		27.	$6100	11%	
9.	$5000		$350	28.	$3100		$93
10.	$1000	5%		29.	$4860	25%	
11.		10%	$66	30.		1%	$47
12.	$850		$102	31.		5%	$220
13.		15%	$123	32.		6%	$126
14.	$600	20%		33.	$5060	15%	
15.	$900	4%		34.		8%	$408
16.	$150	3%		35.	$4100	7%	
17.	$3000		$60	36.	$7500		$825
18.		8%	$648	37.		15%	$306
19.	$4000		$360	38.	$8400		$336

Interest 2

Find the simple interest by writing and simplifying an I=PRT equation.

1. on $1200 at 8% for 7 months	9. on $2600 at 9% for 6 months
2. on $6000 at 6% for 3 months	10. on $70,000 at 5% for 3 months
3. on $2000 at 8% for 9 months	11. on $3600 at 8% for 8 months
4. on $24,000 at 5% for 15 months	12. on $2400 at 6% for 18 months
5. on $6000 at 6% for 3 months	13. on $60,000 at 11% for 2 months
6. on $4800 at 4% for 9 months	14. on $59,400 at 12% for 1 month
7. on $5500 at 9% for 8 months	15. on $7200 at 7% for 15 months
8. on $3200 at 10% for 9 months	16. on $6400 at 10% for 9 months

Interest 3

Complete the chart of compound interest using a calculator given the annual rate.

	PRINCIPAL	RATE	COMPOUND	YEARS	FORMULA	VALUE
1.	$2000	6%	annually	5		
2.	$350	8%	quarterly	2		
3.	$5000	4.8%	monthly	1		
4.	$3000	2%	quarterly	2.5		
5.	$6000	10%	semi-annually	6		
6.	$500	3.6%	semi-monthly	1.5		
7.	$2500	5%	annually	2		
8.	$5500	2.4%	monthly	2		
9.	$600	12%	monthly	1.5		
10.	$1750	9%	quarterly	4		
11.	$1000	12%	semi-annually	7		
12.	$900	6.25%	semi-annually	3		
13.	$8000	4%	quarterly	2		
14.	$250	6%	quarterly	3		
15.	$2600	10%	annually	3		
16.	$1100	15%	semi-monthly	1/2		
17.	$1000	12%	semi-annually	1		
18.	$35,000	8%	annually	3		
19.	$2000	10%	quarterly	1		

Interest 4

Answer by writing and solving an I=PRT equation.

1. Find the principal that yields $198 simple interest at 5.5% for 16 months.

7. Simple annual interest for 42 months on $1300 was $182. Find the rate.

2. Simple annual interest for 4 years on $9000 was $1980. Find the rate.

8. Find the rate that yields $100 simple interest on $500 after 8 years.

3. Find the principal that yields $27 simple interest at 4.5% for 9 months.

9. Simple annual interest for 3.5 years on $3500 was $735. Find the rate.

4. Find the rate that yields $3250 simple interest on $6500 after 5 years.

10. Find the principal that yields $22 simple interest at 5.5% for 8 months.

5. Find the principal that yields $195 simple interest at 6.5% for 10 months.

11. Simple annual interest for 6 years on $6000 was $2880. Find the rate.

6. Simple annual interest paying 6% was $840 after 2 years. Find the principal.

12. Simple annual interest paying 11% was $3080 after 7 years. Find the principal.

Least Common Multiple 1

Find the LCM by mental math.

1. 12 and 60	17. 26 and 39	33. 2, 9, and 20
2. 19 and 20	18. 6, 11, and 12	34. 18 and 24
3. 9 and 12	19. 15 and 80	35. 6, 22, and 33
4. 16 and 24	20. 10 and 18	36. 9 and 22
5. 14 and 77	21. 25 and 40	37. 4, 5, and 10
6. 8, 10, and 12	22. 3, 8, and 16	38. 44 and 121
7. 20 and 21	23. 40 and 56	39. 10, 15, and 25
8. 50 and 60	24. 10 and 21	40. 5, 9, and 10
9. 3, 11, and 66	25. 2, 3, and 4	41. 6, 9, and 15
10. 16 and 20	26. 9, 10, and 11	42. 15 and 40
11. 14 and 21	27. 2, 5, and 13	43. 2, 5, and 6
12. 6, 7, and 10	28. 25 and 35	44. 5, 6, and 21
13. 50 and 80	29. 8, 10, and 20	45. 4, 11, and 16
14. 20 and 36	30. 8, 9, and 10	46. 5, 6, and 10
15. 6, 12, and 18	31. 20 and 39	47. 25 and 325
16. 2, 9, and 11	32. 5, 7, and 20	48. 18, 45, and 90

Least Common Multiple 2

Find the least common multiple by prime factorization.

1.	45	75	9.	168	224	
2.	36	80	10.	135	225	
3.	63	77	11.	108	405	
4.	48	72	12.	126	297	
5.	54	60	13.	21	24	84
6.	27	90	14.	42	52	70
7.	144	216	15.	16	56	98
8.	675	750	16.	40	88	96

Least Common Multiple 3

Find the LCM.

1. $LCM(a^3b^4c^5, a^2b^6c^3)$

2. $LCM(16x^4y^5z^3, 48x^3y^6z^3)$

3. $LCM(5a^3s^4, 35a^4s^3, 10a^2s^3)$

4. $LCM(35a^2b^2c^2, 42a^3b^5)$

5. $LCM(6f^7g^3h^6, 10f^4g^5h^8)$

6. $LCM(36d^5e^2h^6, 144d^3e^6h^8)$

7. $LCM(3x^4y^5z^6, 51x^3y^2z^8)$

8. $LCM(15x^2y^3z^4, 20w^2y^4z^3)$

9. $LCM(4d^2e^3, 5d^4e^4, 6d^6e^5)$

10. $LCM(36p^4q^2r^8, 90p^5q^2r^3)$

11. $LCM(15x^9y^6z^5, 70x^5y^6z^9)$

12. $LCM(16r^4st, 56r^6s^3t^5, 4r^2s^3t)$

13. $LCM(12rst, 16r^2s^5t^4, 8r^3s^3t^6)$

14. $LCM(11a^4b^4c, 14a^9b^3c^6)$

15. $LCM(15a^5c^7, 9a^3b^9)$

16. $LCM(14f^7g^6h^4, 16f^3g^4h^2)$

17. $LCM(5w^7x^4y^5z^3, 7w^2x^5y^4z)$

18. $LCM(6b^3d^5g^6, 18b^4d^5h^2)$

19. $LCM(10p^9q^2r^4, 14p^5q^5r^6)$

20. $LCM(6m^3n^6p^7, 15m^2n^4p^8)$

21. $LCM(6e^3f^5g^5, 8e^6f^5g^7)$

22. $LCM(5g^7s^6, 10g^6s^7, 15g^8s^3)$

23. $LCM(14x^3z^7, 21y^2z^6, 35x^4y^3)$

24. $LCM(8m^8n^2, 12p^9n^5, 20m^3p^4)$

25. $LCM(6a^5b^5, 33a^6b^6c, 22a^5b^6)$

26. $LCM(9r^3s^6, 8pr^6s^3, 12r^5s^5)$

Least Common Multiple 4

Answer as indicated.	*Given one of two numbers m and their GCF and LCM, find the second number n.*
1. A collection of $100 bills may be put in stacks of 2, 5, or 7 with none left over. If the number of bills is between 100 and 200, find the value of the money.	7. m = 550 GCF(m, n) = 2 x 5 x 11 LCM(m, n) = 2^3 x 5^2 x 11^2
2. Two runners start a race at the same time from the same line. One completes a lap every 8 minutes, the other a lap every 10 minutes. How long after they start do they first cross the line together?	8. m = 630 GCF(m, n) = 2 x 5 x 7 LCM(m, n) = 2^2 x 3^2 x 5^4 x 7
3. Find the least number of $50 bills that may be put in stacks of 3, 7, or 11 with none left over. What is the value of the money?	9. m = 520 GCF(m, n) = 2 x 5 x 13 LCM(m, n) = 2^2 x 3 x 5^2 x 7
4. A school that is in session Monday through Friday has an 8-day-cycle schedule of classes that starts on a Tuesday. After how many days is the day of the week again Tuesday and the order of classes matches the first day's?	10. m = 1485 GCF(m, n) = 3 x 5 x 11 LCM(m, n) = 3^3 x 5^2 x 7 x 11
5. A set of marbles in groups of 2s, 3s, or 4s has 1 marble left but 0 left in groups of 5s. Find the number of marbles greater than 25 and less than 100.	11. m = 1540 GCF(m, n) = 2 x 5 x 7 LCM(m, n) = 2^2 x 5^2 x 7^2 x 11
6. A mother is bringing cookies for 1 class, a teacher for another. The mother cannot remember whether the class of 32 or 24 students is hers. Find the least number of cookies to ensure all cookies are given out and each child gets the same number.	12. m = 5390 GCF(m, n) = 2 x 5 x 11 LCM(m, n) = 2^2 x 5^2 x 7^2 x 11^2

Line Segments 1

Find the midpoint or endpoint of the vertical or horizontal segment.

	ENDPOINT	ENDPOINT	MIDPOINT		ENDPOINT	ENDPOINT	MIDPOINT
1.	(6, 6)	(18, 6)		20.	(5, 3)		(5, 11)
2.	(−10, 10)	(1, 10)		21.	(4, −3)		(10.5, −3)
3.	(15, −3)	(18, −3)		22.		(14, 1)	(14, −3)
4.	(7, 13)	(7, −11)		23.		(18, 9)	(18, 4.5)
5.	(−8, 4)	(7, 4)		24.	(12, 3)		(12, −.5)
6.	(0, 9)	(0, 12)		25.	(3, 5)		(5.5, 5)
7.	(9, 5)	(16, 5)		26.		(−2, 1)	(−2, 7)
8.	(11, −2)	(11, 4)		27.		(−5, 2)	(2.5, 2)
9.	(−3, 1)	(19, 1)		28.	(2.9, 2)		(6.4, 2)
10.	(−20, 6)	(13, 6)		29.		(8.9, 9)	(8.5, 9)
11.	(11, 4)	(18, 4)		30.	(35, 36)		(37, 36)
12.	(8, 0)	(−7, 0)		31.		(24, 17)	(24, 12)
13.	(1, 3)	(1, −9)		32.	(−2, 15)		(2, 15)
14.	(0, 8)	(7, 8)		33.	(30, 35)		(35, 35)
15.	(8, 18)	(11, 18)		34.		(9.5, 1)	(12.5, 1)
16.	(20, 16)	(29, 16)		35.	(11, −12)		(11, 1.5)
17.	(6.3, 8)	(1.3, 8)		36.	(15, 1)		(20.5, 1)
18.	(9, 13)	(16, 13)		37.		(3, 8.5)	(3, 2.5)
19.	(6, −10)	(6, 10)		38.		(11, 15)	(11, −5)

Line Segments 2

Find the endpoint or consecutive points of trisection of the vertical or horizontal segment.

	ENDPOINT	TRI-POINT	TRI-POINT	ENDPOINT
1.	(6, 6)			(27, 6)
2.	(–3, 10)			(–3, 46)
3.	(8, 7)		(36, 7)	
4.		(3, 25)		(3, 41)
5.	(4, 9)			(49, 9)
6.	(8, –3)			(8, 24)
7.		(14, 3)		(14, 23)
8.	(2, 13)		(36, 13)	
9.	(–4, 1.5)			(29, 1.5)
10.	(–7, 1)			(–7, 40)
11.	(12, –6)		(–4, –6)	
12.		(5, 38)		(5, 18)
13.	(0, –1)			(0, –16)
14.	(3, 11)			(60, 11)
15.		(–8, –1)		(–8, 13)
16.	(6, 10)			(54, 10)
17.	(–5, 2)			(55, 2)
18.	(–1, 40)			(–1, 10)
19.	(42, 12)		(20, 12)	

Line Segments 3

Find the midpoint of the diagonal segment. | *Find the endpoint of the diagonal segment.*

	ENDPOINT	ENDPOINT	MIDPOINT		ENDPOINT	ENDPOINT	MIDPOINT
1.	(6, –6)	(10, 12)		20.	(13, 30)		(14.5, 36)
2.	(–1, 3)	(7, 6)		21.	(–5, –9)		(–1.5, –5)
3.	(2, 4)	(14, 7)		22.	(12, –6)		(11, –5)
4.	(4, –8)	(–4, 7)		23.	(–20, 7)		(–15, 3)
5.	(–3, 5)	(1, –7)		24.	(15, 9)		(14.5, 1.5)
6.	(9, 5)	(–2, 7)		25.	(5, 16)		(2.5, 10)
7.	(13, –6)	(6, 6)		26.	(8, –10)		(8.5, –9.5)
8.	(6, 2)	(12, –7)		27.	(14, 5)		(8.5, 2)
9.	(8, 7)	(–4, 10)		28.	(4, 20)		(3, 18)
10.	(1, 8)	(11, –6)		29.	(5, 5)		(–3, 9)
11.	(2, 18)	(7, –8)		30.	(0, 9)		(2.5, 4.5)
12.	(3, 5)	(–3, –5)		31.	(8, 7)		(5.5, 1)
13.	(–3, –4)	(–5, –6)		32.	(20, 0)		(16.5, 4)
14.	(11, –3)	(17, –8)		33.	(5, 13)		(11.5, 9)
15.	(3, 12)	(–7, 15)		34.	(30, 2)		(25, 3)
16.	(–4, 11)	(–9, 15)		35.	(5, 5)		(2, 1)
17.	(12, 12)	(22, –4)		36.	(58, 8)		(54, 5)
18.	(–3, –8)	(9, –1)		37.	(–2, 10)		(8, –11)
19.	(14, –12)	(11, 15)		38.	(22, 9)		(14, 13)

210

Line Segments 4

Find the endpoint or consecutive points of trisection of the diagonal segment.

	ENDPOINT	TRI-POINT	TRI-POINT	ENDPOINT
1.	(0, 0)			(21, 21)
2.	(2, 4)			(17, 13)
3.	(1, 18)		(13, 4)	
4.		(−1, 3)		(7, 19)
5.	(9, 3)			(−6, −6)
6.	(7, 2)			(23.5, 29)
7.		(7.6, 3.1)		(0.8, 15.3)
8.	(8, 13)		(−6, 5)	
9.	(−4, 3.5)			(14, 14)
10.	(−7, 6)			(8, −12)
11.	(14, −4)		(8.8, 1)	
12.		(18, 3)		(44, −11)
13.	(4, −1)			(15.1, −19)
14.	(−1, 9)			(−28, 1.5)
15.		(−7.5, −1)		(−10.5, 9)
16.	(3, 7)			(7.2, 14.8)
17.	(−8, 0)			(16, 3.3)
18.	(−1, 20)			(−4, 0.5)
19.	(21, 12)		(26.4, −6)	

MAVA Math: Enhanced Skills Copyright © 2015 Marla Weiss

Line Segments 5

Find the coordinate of the point on the number line.

1.	3/5 of the way from −7 to 13	10.	1/3 of the way from −4 to −2.5
2.	1/4 of the way from 10 to 0	11.	1/6 of the way from −1.2 to 1.2
3.	.6 of the way from −1 to 3	12.	2/3 of the way from −4 to 5
4.	3/4 of the way from 4 to −2	13.	.75 of the way from 4.2 to −2.2
5.	2/3 of the way from −2 to 4	14.	2/3 of the way from .15 to .9
6.	2/5 of the way from 2 to −.5	15.	2/5 of the way from −3 to 2
7.	.4 of the way from 2 to −3	16.	3/4 of the way from −2 to 10
8.	5/6 of the way from −1 to 11	17.	.6 of the way from 9 to −6
9.	5/7 of the way from 7 to −7	18.	7/9 of the way from −9 to 18

Line Segments 6

Answer as indicated.

1. On $\overline{PR}$, PQ = 12 and QR = 8. What is the length of the segment joining the midpoints of $\overline{PQ}$ and $\overline{QR}$?

2. A is the midpoint of $\overline{TH}$, and TA = 12. Points E and C are trisection points of $\overline{TH}$. Find AC.

3. A, B, C, and D are points on a line with D the midpoint of BC. AB = 16. AC = 4. Find all possible values of AD.

4. E divides $\overline{DF}$ into $\overline{DE}$ and $\overline{EF}$ in the ratio 5:2. DF = 21. Find EF.

5. AC = DF. AB = EF. Find BC − DE.

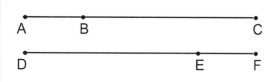

A   B   C

D   E   F

6. R and Q are on $\overline{PS}$ with PS = 36. QS = 3PQ. QR = 2RS. Find RS.

7. A, B, C, and D lie on a segment in that order. AB:BC = 4:5, and BC:CD = 7:2. AB = 14. Find AD.

8. B and C are on $\overline{AD}$ with AD = 36. BD is twice AB, and CD is one third BC. Find CD.

9. E, F, G, and H are on a segment. H is the midpoint of $\overline{FG}$. EF = 60. EG = 12. FG = 72. Find EH.

10. GT = 70. R is the midpoint of $\overline{GE}$. E is the midpoint of $\overline{RA}$. A is the midpoint of $\overline{RT}$. Find RA.

11. A and B are each placed to the right of C on a segment such that 3BC = 5AC. Find AB:AC.

12. P, Q, R, S, and T lie on a segment in that order. R is a midpoint of $\overline{QS}$. QS = 9. RT = 12. PT = 22.5. Find PQ.

Logic 1

Prove the statement by evaluating the logical equivalency of both sides of the equation.

1. A AND (B OR C) = (A AND B) OR (A AND C)   Distributive Property

A	B	C	X=B OR C	A AND X	Y=A AND B	Z=A AND C	Y OR Z
TRUE	TRUE	TRUE					
TRUE	TRUE	FALSE					
TRUE	FALSE	TRUE					
TRUE	FALSE	FALSE					
FALSE	TRUE	TRUE					
FALSE	TRUE	FALSE					
FALSE	FALSE	TRUE					
FALSE	FALSE	FALSE					

2. A OR (B AND C) = (A OR B) AND (A OR C)   Distributive Property

A	B	C	X=B AND C	A OR X	Y=A OR B	Z=A OR C	Y AND Z
TRUE	TRUE	TRUE					
TRUE	TRUE	FALSE					
TRUE	FALSE	TRUE					
TRUE	FALSE	FALSE					
FALSE	TRUE	TRUE					
FALSE	TRUE	FALSE					
FALSE	FALSE	TRUE					
FALSE	FALSE	FALSE					

3. NOT (A OR B) = (NOT A) AND (NOT B)   DeMorgan's Law

A	B		X=A OR B	NOT X	Y = NOT A	Z = NOT B	Y AND Z
TRUE	TRUE						
TRUE	FALSE						
FALSE	TRUE						
FALSE	FALSE						

4. NOT (A AND B) = (NOT A) OR (NOT B)   DeMorgan's Law

A	B		X=A AND B	NOT X	Y=NOT A	Z=NOT B	Y OR Z
TRUE	TRUE						
TRUE	FALSE						
FALSE	TRUE						
FALSE	FALSE						

Logic 2

Write the converse, inverse, and contrapositive of the true implication. Label TRUE or FALSE.

1. If I live in Reno, then I live in Nevada.

Converse:

Inverse:

Contrapositive:

2. If a triangle has two congruent sides, then the angles opposite the sides are congruent.

Converse:

Inverse:

Contrapositive:

3. If two lines in a plane are parallel, then the alternate interior angles are congruent.

Converse:

Inverse:

Contrapositive:

4. If two lines in a plane are perpendicular, then the lines have negative reciprocal slopes.

Converse:

Inverse:

Contrapositive:

5. If a median, altitude, and angle bisector of a triangle are all the same line segment, then the triangle is isosceles.

Converse:

Inverse:

Contrapositive:

Matrices 1

Add or subtract as indicated.

1. $\begin{bmatrix} 17 & -5 & 33 \\ 16 & 17 & -42 \end{bmatrix} + \begin{bmatrix} -3 & 11 & 44 \\ -4 & -9 & 57 \end{bmatrix} + \begin{bmatrix} 21 & -6 & -9 \\ 24 & 53 & 42 \end{bmatrix}$

2. $\begin{bmatrix} -5 & 65 \\ 14 & -23 \end{bmatrix} + \begin{bmatrix} 17 & 15 \\ 16 & -12 \end{bmatrix} - \begin{bmatrix} -4 & 24 \\ 13 & -10 \end{bmatrix} + \begin{bmatrix} 12 & 11 \\ 10 & -13 \end{bmatrix}$

3. $\begin{bmatrix} 10 & 45 \\ 13 & -22 \\ 56 & -30 \end{bmatrix} - \begin{bmatrix} 23 & 25 \\ -7 & -40 \\ 16 & -60 \end{bmatrix} - \begin{bmatrix} 12 & 24 \\ 16 & -18 \\ 28 & -50 \end{bmatrix} + \begin{bmatrix} 50 & 12 \\ 24 & -10 \\ 46 & -20 \end{bmatrix}$

4. $\begin{bmatrix} -7 & -1 \\ 17 & 20 \\ 16 & -42 \\ 29 & 14 \end{bmatrix} - \begin{bmatrix} -11 & 45 \\ 17 & -25 \\ -13 & -42 \\ 18 & -15 \end{bmatrix} + \begin{bmatrix} -13 & 22 \\ 17 & -35 \\ 13 & 42 \\ 21 & -11 \end{bmatrix} + \begin{bmatrix} 9 & 24 \\ 17 & -31 \\ 13 & -42 \\ -5 & 38 \end{bmatrix}$

5. $\begin{bmatrix} -8 & 17 & 35 \\ 39 & 21 & -22 \end{bmatrix} + \begin{bmatrix} -9 & 37 & 35 \\ 16 & -5 & -22 \end{bmatrix} + \begin{bmatrix} -11 & 33 & 25 \\ 44 & 16 & 40 \end{bmatrix}$

6. $\begin{bmatrix} -5 & -4 & 18 \\ 35 & 16 & -29 \end{bmatrix} + \begin{bmatrix} 25 & 19 & 21 \\ -8 & 16 & -31 \end{bmatrix} + \begin{bmatrix} -6 & -3 & 19 \\ 55 & 16 & -11 \end{bmatrix}$

7. $\begin{bmatrix} 29 & 9 \\ 13 & -34 \\ 14 & -42 \end{bmatrix} - \begin{bmatrix} -5 & 20 \\ 20 & -40 \\ 16 & -42 \end{bmatrix} + \begin{bmatrix} -11 & 52 \\ 27 & -12 \\ 12 & -43 \end{bmatrix} + \begin{bmatrix} -11 & 11 \\ 14 & -4 \\ 38 & 50 \end{bmatrix}$

8. $\begin{bmatrix} 50 & 15 \\ 36 & -21 \\ 52 & -33 \end{bmatrix} + \begin{bmatrix} -9 & 27 \\ 34 & -12 \\ 28 & -17 \end{bmatrix} - \begin{bmatrix} -11 & -5 \\ 20 & -11 \\ -8 & -22 \end{bmatrix} - \begin{bmatrix} -5 & 17 \\ 13 & -10 \\ 12 & -18 \end{bmatrix}$

Matrices 2

Multiply by the scalar using mental math.

1.
$$5 \begin{bmatrix} 12 & -3 & 10 \\ 15 & 13 & 11 \end{bmatrix}$$

9.
$$2 \begin{bmatrix} 2.9 & 3.6 & 4.6 \\ 1.7 & 3.9 & 3.8 \end{bmatrix}$$

2.
$$4 \begin{bmatrix} -6 & 14 \\ 40 & -17 \end{bmatrix}$$

10.
$$6 \begin{bmatrix} -7 & 19 \\ 45 & -25 \end{bmatrix}$$

3.
$$11 \begin{bmatrix} 17 & 31 \\ 27 & 87 \\ 76 & 54 \end{bmatrix}$$

11.
$$25 \begin{bmatrix} 11 & -30 \\ 25 & -15 \\ 19 & -12 \end{bmatrix}$$

4.
$$12 \begin{bmatrix} 20 & 11 \\ 8.5 & -5 \\ 12 & 15 \end{bmatrix}$$

12.
$$15 \begin{bmatrix} 60 & 15 \\ 2.5 & -7 \\ 11 & 16 \end{bmatrix}$$

5.
$$1.5 \begin{bmatrix} -6 & 32 & 60 \\ 50 & 24 & -8 \end{bmatrix}$$

13.
$$1.1 \begin{bmatrix} -33 & 45 \\ -1.3 & 32 \end{bmatrix}$$

6.
$$-6 \begin{bmatrix} -8 & 12 & -9 \\ 1 & -11 & 0 \end{bmatrix}$$

14.
$$-9 \begin{bmatrix} 31 & -25 \\ 14 & -72 \end{bmatrix}$$

7.
$$20 \begin{bmatrix} -9 & 35 \\ 13 & -7 \\ 22 & 46 \end{bmatrix}$$

15.
$$30 \begin{bmatrix} -9 & 111 \\ 99 & -75 \\ 80 & 151 \end{bmatrix}$$

8.
$$10 \begin{bmatrix} 6.5 & 2.36 \\ .92 & .345 \\ 5.1 & 89.2 \end{bmatrix}$$

16.
$$50 \begin{bmatrix} 8.2 & .12 \\ .5 & 2.4 \\ 3.5 & 10.1 \end{bmatrix}$$

Matrices 3

Multiply the matrices.

1.
$$\begin{bmatrix} 6 & 9 \\ 3 & 7 \end{bmatrix} \begin{bmatrix} 5 & 8 \\ 4 & 2 \end{bmatrix}$$

2.
$$\begin{bmatrix} 9 & -8 \\ -5 & 6 \\ -4 & 3 \\ -3 & 2 \end{bmatrix} \begin{bmatrix} 11 & -2 \\ 10 & -3 \end{bmatrix}$$

3.
$$\begin{bmatrix} 100 & -25 & 125 \\ 200 & -15 & 300 \end{bmatrix} \begin{bmatrix} -1 & 2 \\ 2 & -4 \\ 3 & 1 \end{bmatrix}$$

4.
$$\begin{bmatrix} 80 & 20 \\ 40 & 10 \\ 50 & 70 \\ 60 & 90 \end{bmatrix} \begin{bmatrix} 1.5 & 3.5 & 5.5 \\ 2.5 & 4.5 & 6.5 \end{bmatrix}$$

5.
$$\begin{bmatrix} 5 & -9 & 11 \\ 6 & -5 & 7 \end{bmatrix} \begin{bmatrix} 6 & 11 \\ 5 & 10 \\ 7 & 12 \end{bmatrix}$$

6.
$$\begin{bmatrix} -7 & -6 & -8 \\ -4 & -2 & -3 \end{bmatrix} \begin{bmatrix} -5 & -6 \\ 10 & -1 \\ -9 & 11 \end{bmatrix}$$

7.
$$\begin{bmatrix} 1 & -2 & 3 \\ 4 & 0 & -5 \\ 6 & -7 & 8 \end{bmatrix} \begin{bmatrix} 11 & 10 & 9 \\ 8 & 7 & -6 \\ 5 & -4 & 3 \end{bmatrix}$$

8.
$$\begin{bmatrix} 10 & -5 & 11 \\ 30 & 12 & 25 \\ 15 & -2 & 20 \end{bmatrix} \begin{bmatrix} 16 & 14 & 20 \\ 15 & 10 & -3 \\ 11 & 16 & 12 \end{bmatrix}$$

Matrices 4

Solve for the variables.

1. $\begin{bmatrix} 3x + 1 & 33 \\ 5y - 4 & 42 \end{bmatrix} = \begin{bmatrix} 52 & 3w + 3 \\ -64 & 4z - 6 \end{bmatrix}$

8. $\begin{bmatrix} 2x + 4 & 70 \\ 3y - 6 & 31 \end{bmatrix} = \begin{bmatrix} 47 & 9w + 7 \\ 42 & 3z - 8 \end{bmatrix}$

2. $\begin{bmatrix} 7w - 2 & 65 \\ 5z - 5 & 35 \end{bmatrix} = \begin{bmatrix} 12 & 3x + 8 \\ 15 & 6y - 7 \end{bmatrix}$

9. $\begin{bmatrix} 75 & 4x + 1 \\ 9y - 6 & 42 \end{bmatrix} = \begin{bmatrix} 7w + 5 & 89 \\ 66 & 4z - 6 \end{bmatrix}$

3. $\begin{bmatrix} 24 & 5x + 7 \\ 4y - 3 & 41 \end{bmatrix} = \begin{bmatrix} 3w + 6 & 77 \\ 45 & 7z + 6 \end{bmatrix}$

10. $\begin{bmatrix} 8a + 2 & 50 \\ 7c + 6 & 68 \end{bmatrix} = \begin{bmatrix} 90 & 9b + 5 \\ -8 & 7d - 9 \end{bmatrix}$

4. $\begin{bmatrix} 9x + 2 & 28 \\ 5y + 1 & 38 \end{bmatrix} = \begin{bmatrix} 56 & 3w - 2 \\ -4 & 3z + 5 \end{bmatrix}$

11. $\begin{bmatrix} 3x + 5 & 38 \\ 9y - 9 & 80 \end{bmatrix} = \begin{bmatrix} 80 & 9w + 2 \\ 90 & 4z - 4 \end{bmatrix}$

5. $\begin{bmatrix} 7x + 1 & 43 \\ 4y + 9 & 53 \end{bmatrix} = \begin{bmatrix} -34 & 8w + 3 \\ 49 & 5z - 7 \end{bmatrix}$

12. $\begin{bmatrix} 6x + 9 & 33 \\ 5z - 4 & 50 \end{bmatrix} = \begin{bmatrix} -21 & 6w + 3 \\ 66 & 9y - 4 \end{bmatrix}$

6. $\begin{bmatrix} 2a + 9 & 32 \\ 7c + 6 & 55 \end{bmatrix} = \begin{bmatrix} -11 & 5b + 7 \\ 69 & 8d - 9 \end{bmatrix}$

13. $\begin{bmatrix} 4a + 1 & 57 \\ 3c + 4 & 41 \end{bmatrix} = \begin{bmatrix} -51 & 6b - 3 \\ 52 & 7d - 8 \end{bmatrix}$

7. $\begin{bmatrix} 69 & 7x + 5 \\ 7y + 8 & 45 \end{bmatrix} = \begin{bmatrix} 6w + 3 & 40 \\ -6 & 3z - 6 \end{bmatrix}$

14. $\begin{bmatrix} 43 & 5x + 7 \\ 8y - 3 & 66 \end{bmatrix} = \begin{bmatrix} 6w + 1 & 47 \\ 61 & 5z - 9 \end{bmatrix}$

Measurement 1

Convert the measurements in the standard system.

1. 2.5 gallon = _____ quarts	20. 3.5 pints = _____ cups
2. 5.5 quarts = _____ pints	21. 4/3 yards = _____ inches
3. 6.5 dozen = _____	22. 2.75 ton = _____ pounds
4. 3/4 quart = _____ cups	23. 1.4 ton = _____ pounds
5. 5/4 yard = _____ inches	24. 1/4 ton = _____ ounces
6. 1/4 gallon = _____ pints	25. 5/4 pound = _____ ounces
7. 7/4 feet = _____ inches	26. 4/3 dozen = _____
8. 3.25 yard = _____ feet	27. 3/4 gallon = _____ quarts
9. 2.75 pound = _____ ounces	28. 2/5 ton = _____ pounds
10. 3.1 tons = _____ pounds	29. .375 quart = _____ cups
11. .125 gallon = _____ cups	30. .3 mile = _____ feet
12. .6 ton = _____ pounds	31. 1.125 quart = _____ cups
13. 1.75 dozen = _____	32. .001 ton = _____ pounds
14. 8/3 yard = _____ inches	33. 5/9 yard = _____ inches
15. 2.6 ton = _____ pounds	34. 1.5 pints = _____ cups
16. 1/4 mile = _____ feet	35. .875 gallon = _____ quarts
17. 7/2 pound = _____ ounces	36. .05 ton = _____ pounds
18. 19.5 feet = _____ yards	37. 54 inches = _____ yards
19. 7/6 dozen = _____	38. 40 ounces = _____ pounds

Measurement 2

Operate and regroup.

1.   3 mi   101 yd   2 ft   5 in
   + 1 mi   200 yd   9 ft   8 in
		9.
		      7 yd 1 ft 2 in
		   − 3 yd 2 ft 7 in

2. 6 · (3 lb 7 oz)		10. (8 yd 2 ft 8 in) ÷ 5

3.
      6 yd   1 ft   1 in
   − 2 yd   2 ft   8 in
		11. (8 yd 2 ft 6 in) ÷ 3

4.
      15 days   6 hr   7 min
   − 5 days   17 hr   57 min
		12. 4 · (2 yd 2 ft 7 in)

5.   5 tons   560 lb   17 oz
   + 2 tons   1950 lb   15 oz
		13. 2 mi 750 yd 74 ft 7 in
		   + 3 mi 1100 yd 70 ft 9 in

6. 4 · (5 hr 20 min 40 sec)		14.
		      8 days 5 hr 3 min 5 sec
		   − 4 days 5 hr 9 min 17 sec

7. $\frac{1}{3}$ (7 days 8 hr 9 min)		15.
		      10 yd
		   − 3 yd 1 ft 8 in

8. 7 · (3 yd 3 ft 8 in)		16. $\frac{1}{4}$ (5 hr 19 min 4 sec)

Measurement 3

Answer as indicated.

1. 1 cm + 1 m = _____ m	12. 1.5 L – 150 mL = _____ mL
2. 1 cm + 1 m = _____ cm	13. 6 cm + 7 dm = _____ dm
3. 1 mm + 1 m = _____ mm	14. 20 dm + 8 cm = _____ cm
4. 1 dm + 1 km = _____ m	15. 6.1 km + 4.1 m = _____ m
5. 1 m + 1 dm = _____ cm	16. 3 cm + 2 m = _____ mm
6. 1 cm + 1 dm = _____ m	17. 2 kg + 2 g = _____ g
7. 1 cm + 1 mm = _____ m	18. 5 g – 200 mg = _____ mg
8. 1 cm + 1 mm = _____ dm	19. 3.5 L + 1.5 mL = _____ mL
9. 1 cm + 1 mm = _____ cm	20. 0.6 L + 100 mL = _____ L
10. 1 cm + 1 km = _____ m	21. 1.5 km + 0.5 m = _____ m
11. 1 dm + 1 mm = _____ cm	22. 0.5 km + 2 m = _____ km

Measurement 4

Answer as indicated.

1.	10 sq yd = _____ sq ft	12.	321 ft by 30 ft = _____ sq yd
2.	10 sq yd = _____ sq in	13.	144 in by 120 in = _____ sq ft
3.	120 sq ft = _____ sq in	14.	108 in by 180 in = _____ sq yd
4.	10 cu ft = _____ cu in	15.	11 yd by 9 yd = _____ sq ft
5.	2 sq yd = _____ sq in	16.	5 ft by 2 ft = _____ sq in
6.	3 sq ft = _____ sq in	17.	111 ft by 33 ft by 21 ft = _____ cu yd
7.	3 cu yd = _____ cu ft	18.	36 in by 72 in by 96 in = _____ cu ft
8.	5 sq ft = _____ sq in	19.	54 in by 36 in by 72 in = _____ cu yd
9.	2 sq yd = _____ sq ft	20.	5 ft by 10 ft by 2 ft = _____ cu in
10.	10 cu yd = _____ cu ft	21.	15 ft by 12 yd by 48 in = _____ cu yd
11.	5 cu ft = _____ cu in	22.	90 in by 18 yd by 66 ft = _____ cu yd

Mental Math 1

Multiply mentally using the rule for 11.

1. 81 x 11	17. 54 x 11	33. 77 x 11	49. 38 x 11
2. 33 x 11	18. 14 x 11	34. 11 x 55	50. 46 x 11
3. 22 x 11	19. 90 x 11	35. 48 x 11	51. 56 x 11
4. 63 x 11	20. 11 x 43	36. 69 x 11	52. 88 x 11
5. 16 x 11	21. 11 x 61	37. 89 x 11	53. 75 x 11
6. 51 x 11	22. 11 x 23	38. 95 x 11	54. 19 x 11
7. 40 x 11	23. 72 x 11	39. 47 x 11	55. 37 x 11
8. 11 x 52	24. 17 x 11	40. 39 x 11	56. 76 x 11
9. 26 x 11	25. 32 x 11	41. 29 x 11	57. 68 x 11
10. 44 x 11	26. 53 x 11	42. 11 x 57	58. 11 x 49
11. 11 x 15	27. 34 x 11	43. 74 x 11	59. 99 x 11
12. 53 x 11	28. 11 x 21	44. 67 x 11	60. 11 x 78
13. 11 x 24	29. 25 x 11	45. 59 x 11	61. 79 x 11
14. 11 x 71	30. 13 x 11	46. 11 x 64	62. 66 x 11
15. 62 x 11	31. 36 x 11	47. 58 x 11	63. 28 x 11
16. 35 x 11	32. 42 x 11	48. 65 x 11	64. 11 x 73

Mental Math 2

Multiply mentally, pulling out factors of 10.

1. 40 x 30 x 700	17. 770 x 50 x 20	33. 120 x 60 x 50
2. 1600 x 300 x 200	18. 20 x 450 x 100	34. 100 x 190 x 300
3. 20 x 90 x 20	19. 2150 x 200 x 100	35. 2900 x 20 x 10
4. 500 x 70 x 50	20. 50 x 60 x 50 x 60	36. 50 x 300 x 200
5. 140 x 4000	21. 20 x 30 x 40 x 50	37. 590 x 100 x 20
6. 110 x 40 x 30	22. 250 x 20 x 300	38. 270 x 200 x 30
7. 160 x 20 x 20	23. 260 x 30 x 100	39. 200 x 130 x 30
8. 80 x 200 x 80	24. 20 x 60 x 400	40. 60 x 600 x 20
9. 150 x 60 x 20	25. 10 x 220 x 400	41. 70 x 700 x 20
10. 170 x 400 x 10	26. 140 x 30 x 20	42. 80 x 10 x 5000
11. 10 x 1450 x 20	27. 130 x 40 x 20	43. 5500 x 200 x 30
12. 330 x 100 x 20	28. 40 x 60 x 2000	44. 10 x 3520 x 20
13. 20 x 40 x 90	29. 70 x 20 x 50 x 10	45. 30 x 800 x 20
14. 60 x 30 x 50	30. 250 x 900 x 20	46. 150 x 3000 x 40
15. 80 x 400 x 30	31. 350 x 30 x 20	47. 200 x 9000 x 30
16. 150 x 40 x 200	32. 370 x 20 x 200	48. 90 x 70 x 200

Mental Math 3

Operate by mental math.

1. 6012 ÷ 3	17. 872 + 227	33. 46,036 ÷ 2
2. 1775 + 9125	18. 1253 + 8625	34. 7000 − 1620
3. 1719 x 3	19. 2842 ÷ 14	35. 6000 − 2370
4. 641,638 ÷ 2	20. 2613 ÷ 13	36. 1312 x 7
5. 25 x 19	21. 9999 ÷ 11	37. 4842 ÷ 6
6. 8000 − 1650	22. 5105 ÷ 5	38. 1415 x 5
7. 9775 − 1825	23. 6789 − 4321	39. 1315 x 6
8. 35,028 ÷ 7	24. 6045 ÷ 5	40. 36,168 ÷ 4
9. 3624 ÷ 12	25. 9250 − 8725	41. 701,015 ÷ 5
10. 4530 ÷ 15	26. 815 x 4	42. 8725 + 1270
11. 1418 x 3	27. 241,554 ÷ 3	43. 9065 − 7255
12. 42,018 ÷ 6	28. 1765 + 1815	44. 6090 ÷ 15
13. 4000 − 2380	29. 9876 − 1234	45. 8430 + 1570
14. 7590 ÷ 15	30. 4880 ÷ 16	46. 777,777 ÷ 11
15. 1312 x 6	31. 1213 x 7	47. 1211 x 9
16. 324,072 ÷ 8	32. 543,624 ÷ 6	48. 526,036 ÷ 4

Mental Math 4

Multiply mentally using DPMA.

1. 18 x 5	17. 24 x 7	33. 59 x 5
2. 16 x 7	18. 39 x 8	34. 64 x 3
3. 17 x 4	19. 23 x 6	35. 33 x 6
4. 15 x 5	20. 35 x 6	36. 62 x 9
5. 58 x 8	21. 29 x 7	37. 37 x 4
6. 42 x 9	22. 19 x 8	38. 54 x 7
7. 57 x 6	23. 33 x 7	39. 16 x 6
8. 65 x 3	24. 27 x 5	40. 29 x 4
9. 14 x 8	25. 13 x 4	41. 55 x 8
10. 39 x 4	26. 17 x 7	42. 56 x 5
11. 15 x 8	27. 34 x 4	43. 39 x 3
12. 19 x 5	28. 19 x 9	44. 22 x 8
13. 17 x 9	29. 56 x 4	45. 26 x 3
14. 13 x 8	30. 14 x 3	46. 45 x 7
15. 19 x 6	31. 48 x 8	47. 39 x 9
16. 23 x 9	32. 46 x 6	48. 79 x 6

227

Mental Math 5

Multiply or divide mentally by the power of 10.

1. 3.143 x 10	17. 5.632 ÷ 100	33. 5.8 ÷ 1000
2. 3.145 ÷ 100	18. 23.32 x 1000	34. 235.8 ÷ 100
3. 0.2976 x 1000	19. 0.78 x 100	35. 7.442 x 100
4. 4.751 x 100	20. 8.9 ÷ 1000	36. 8.176 ÷ 10
5. 2.31 ÷ 100	21. 0.3276 ÷ 10	37. 5.187 x 1000
6. 35.78 x 1000	22. 4.21 x 100,000	38. 78.99 ÷ 100
7. 643.89 ÷ 100	23. 0.9345 x 100	39. 532.77 x 1000
8. 5.33 ÷ 1000	24. 236.7 x 1000	40. 56.333 x 100
9. 9.544 x 100	25. 0.543 ÷ 10	41. 78.8 x 10,000
10. 82.3 ÷ 1000	26. 643.1 ÷ 100	42. 5.1 x 100,000
11. 306.7 ÷ 10,000	27. 0.22 ÷ 10	43. 2356.77 ÷ 10,000
12. 876.9 ÷ 100	28. 9.11 x 1000	44. 7.324 x 10,000
13. 123.45 x 10	29. 25.75 x 10	45. 21.88 ÷ 1000
14. 56.72 x 100	30. 68.7 x 100	46. 0.832 x 100
15. 2.15 x 10,000	31. 0.5 ÷ 100	47. 431.9 ÷ 10
16. 43.99 ÷ 1000	32. 3.1 ÷ 10	48. 0.7 x 1000

Mental Math 6

Operate mentally using compensation.

1. 998 x 4	12. 92 x 8	23. 3355 + 997
2. 9375 – 97	13. 398 + 97	24. 996 x 5
3. 999 x 48	14. 12 x 999	25. 52 x 999
4. 996 + 8117	15. 94 x 3	26. 6545 – 999
5. 99 x 17	16. 3929 – 996	27. 99 x 13
6. 2985 – 98	17. 96 x 9	28. 4356 – 98
7. 97 x 3	18. 998 + 281	29. 102 x 41
8. 8463 + 997	19. 95 x 7	30. 5495 + 999
9. 95 x 9	20. 70 x 99	31. 4 x 9999
10. 7469 + 999	21. 8265 – 99	32. 6109 – 997
11. 8 x 999	22. 97 x 8	33. 997 x 3

Mental Math 7

Multiply by 101, 1001, or 10001 mentally using DPMA.

1. 34 x 101	17. 89 x 101	33. 101 x 49
2. 56 x 1001	18. 35 x 10,001	34. 31 x 1001
3. 101 x 23	19. 76 x 1001	35. 60 x 101
4. 547 x 1001	20. 192 x 10,001	36. 996 x 1001
5. 756 x 10,001	21. 1224 x 10,001	37. 246 x 10,001
6. 27 x 101	22. 94 x 101	38. 874 x 1001
7. 99 x 10,001	23. 513 x 1001	39. 635 x 1001
8. 1001 x 79	24. 57 x 10,001	40. 97 x 101
9. 52 x 101	25. 78 x 101	41. 99 x 101
10. 295 x 1001	26. 926 x 1001	42. 1001 x 141
11. 25 x 101	27. 72 x 101	43. 10,001 x 478
12. 125 x 1001	28. 1001 x 379	44. 723 x 1001
13. 1234 x 10,001	29. 70 x 1001	45. 28 x 101
14. 41 x 101	30. 22 x 101	46. 75 x 101
15. 82 x 1001	31. 137 x 1001	47. 54 x 1001
16. 65 x 10,001	32. 624 x 10,001	48. 1923 x 10,001

Mental Math 8

Multiply using DPMA.

1. 27 x 12	12. 41 x 23	23. 25 x 47
2. 52 x 13	13. 17 x 32	24. 80 x 14
3. 41 x 15	14. 65 x 23	25. 85 x 21
4. 90 x 15	15. 37 x 21	26. 33 x 15
5. 75 x 22	16. 55 x 31	27. 45 x 32
6. 18 x 12	17. 31 x 26	28. 25 x 38
7. 51 x 23	18. 35 x 16	29. 45 x 26
8. 31 x 14	19. 23 x 32	30. 18 x 14
9. 42 x 41	20. 85 x 42	31. 95 x 24
10. 75 x 43	21. 39 x 21	32. 97 x 12
11. 35 x 42	22. 55 x 42	33. 75 x 26

Mental Math 9

Operate efficiently using APA and CPA.

1.	32 + 7 + 59 + 8 + 6 + 25 + 83 + 44 + 1	12.	53 + 66 + 95 + 21 + 19 + 15 − 43 + 24
2.	9 + 23 + 6 + 51 + 28 + 15 + 7 + 34 + 2	13.	89 + 63 + 72 + 80 + 38 − 44 + 77 + 25
3.	69 + 53 + 66 + 48 + 67 + 71 + 32 + 44	14.	119 + 346 + 153 + 427 + 154 + 331
4.	21 + 76 + 52 + 79 + 87 + 54 + 33 + 18	15.	51 + 65 + 142 + 95 + 9 + 27 + 28 − 87
5.	17 + 81 + 74 + 92 + 83 + 88 + 96 + 99	16.	532 − 24 + 257 + 74 + 20 + 133 + 138
6.	63 + 58 + 86 + 29 + 24 + 91 + 37 + 72	17.	406 + 52 + 173 + 54 + 218 + 87 + 75
7.	77 + 82 + 26 + 41 + 23 + 64 + 39 + 28	18.	64 − 77 + 49 − 53 + 27 + 41 − 84 + 93
8.	84 + 93 + 57 + 89 + 22 + 36 + 98 + 31	19.	105 + 501 + 13 + 149 + 327 + 90 − 55
9.	62 + 94 + 49 + 38 + 73 + 16 + 61 + 97	20.	215 + 342 + 107 + 425 + 333 + 118
10.	78 + 59 + 47 + 25 + 51 + 85 + 12 + 13	21.	95 + 26 + 32 + 77 + 88 + 54 + 43 − 15
11.	27 + 65 + 19 + 34 + 56 + 75 + 43 + 11	22.	72 + 89 + 17 − 85 + 55 − 22 − 29 + 73

Mental Math 10

Operate by mental math.

1. 567 x 5	17. 7461 x 5	33. 4429 x 5
2. 7288 x 5	18. 3665 x 5	34. 7656 x 5
3. 6615 x 5	19. 4656 x 5	35. 3447 x 5
4. 4813 x 5	20. 5025 x 5	36. 983 x 5
5. 389 x 5	21. 5423 x 5	37. 2654 x 5
6. 3662 x 5	22. 1688 x 5	38. 8894 x 5
7. 8692 x 5	23. 487 x 5	39. 9065 x 5
8. 16,854 x 5	24. 1846 x 5	40. 964 x 5
9. 925 x 5	25. 5436 x 5	41. 7292 x 5
10. 6672 x 5	26. 926 x 5	42. 3872 x 5
11. 983 x 5	27. 18,662 x 5	43. 3065 x 5
12. 745 x 5	28. 7845 x 5	44. 5678 x 5
13. 2847 x 5	29. 526 x 5	45. 18,463 x 5
14. 589 x 5	30. 6855 x 5	46. 9256 x 5
15. 9078 x 5	31. 9445 x 5	47. 3229 x 5
16. 643 x 5	32. 2889 x 5	48. 245 x 5

Mental Math 11

Operate as follows: for #1–8 think of a circle; for #9–16 think of quarters; for #17–24 think of a deck of cards or a clock; for #25–32 think of a calendar; for #33–40 think of polygons; for #41–48 think of palindromes.

1. $360 \div 8$	17. 13×4	33. 108×5
2. 72×5	18. $3 \times 4 + 10 \times 4$	34. $45 + 45 + 90$
3. $360 \div 15$	19. $.25 \times 52$	35. 180×5
4. $.25 \times 360$	20. $(60)(2/3)$	36. 60×3
5. $360 \div 40$	21. 12×5	37. 90×4
6. $360 \div 18$	22. 60×60	38. $12 \times 9 / 2$
7. 60×6	23. $52 \div 13$	39. $1260 \div 7$
8. 2×180	24. $60 \div 4$	40. 180×6
9. 25×16	25. $365 \div 7$	41. $3669663 \div 3$
10. 7×25	26. $7 \times 31 + 4 \times 30 + 29$	42. $98689 - 12321$
11. 25×32	27. $.25 \times 12$	43. $212 + 424$
12. 23×25	28. $365 \times 4 + 1$	44. $8558 - 4114$
13. 17×25	29. 52×7	45. $3443 + 1221$
14. 9×25	30. $365 \div 12$	46. $989 - 454$
15. 25×26	31. $31 + 28 + 31$	47. $6776 - 2332$
16. 12×25	32. $365 \div 4$	48. $5335 + 2552$

MAVA Math: Enhanced Skills   Copyright © 2015 Marla Weiss

Mental Math 12

Operate by factoring (DPMA backwards).

1. $$\dfrac{258^2 - 258}{258}$$	11. $$\dfrac{3^4 - 3^2 + 3}{3}$$
2. 12.5% of 52 − 12.5% of 4	12. 497 x 784 + 503 x 784
3. 1378 x 56 + 1378 x 44	13. 173 x 85 − 85 x 52 + 79 x 85
4. 689 x 499 + 311 x 499	14. 5% of 77 − 5% of 32 + 5% of 45
5. 0.5 of 71 + 0.5 of 39	15. 87 x 7145 + 7145 x 13
6. $$\dfrac{175^2 + 350}{175}$$	16. $$\dfrac{343^2 + 343}{343}$$
7. 1536 x 78 + 1536 x 22	17. 393 x 65 + 65 x 47 − 40 x 65
8. 0.75 x 99 − 0.75 x 15	18. 62.5% of 908 − 62.5% of 108
9. $$\dfrac{25^3 - 25^2 + 75}{25}$$	19. $$\dfrac{15^3 - 15^2 + 30}{15}$$
10. 0.25 x 157 − 0.25 x 33	20. half of 357 + half of 143

Mental Math 13

Multiply mentally using the rule for 11.

1. 235 x 11	17. 11 x 7152	33. 11 x 23,625
2. 363 x 11	18. 6324 x 11	34. 40,817 x 11
3. 451 x 11	19. 11 x 1253	35. 57,391 x 11
4. 526 x 11	20. 1632 x 11	36. 43,625 x 11
5. 716 x 11	21. 5436 x 11	37. 85,157 x 11
6. 11 x 815	22. 6145 x 11	38. 57,248 x 11
7. 352 x 11	23. 4445 x 11	39. 35,451 x 11
8. 11 x 534	24. 5674 x 11	40. 11 x 29,415
9. 262 x 11	25. 2953 x 11	41. 11,632 x 11
10. 436 x 11	26. 4652 x 11	42. 66,743 x 11
11. 11 x 624	27. 11 x 2396	43. 72,343 x 11
12. 814 x 11	28. 8546 x 11	44. 11 x 35,814
13. 11 x 636	29. 4539 x 11	45. 48,365 x 11
14. 11 x 271	30. 7683 x 11	46. 66,327 x 11
15. 345 x 11	31. 3939 x 11	47. 27,248 x 11
16. 11 x 535	32. 6758 x 11	48. 17,962 x 11

Mental Math 14

Multiply using the difference of two squares method.

1. 53 x 47	12. 31 x 29
2. 603 x 597	13. 137 x 123
3. 41 x 39	14. 76 x 84
4. 66 x 74	15. 109 x 91
5. 67 x 53	16. 153 x 147
6. 95 x 105	17. 37 x 23
7. 36 x 44	18. 85 x 95
8. 409 x 391	19. 204 x 196
9. 91 x 89	20. 55 x 65
10. 32 x 28	21. 98 x 102
11. 122 x 118	22. 294 x 306

Mental Math 15

Multiply or divide by 4 as indicated.

1. 65 x 4	17. 960 ÷ 4	33. 135 x 4	49. 674 ÷ 4
2. 46 x 4	18. 780 ÷ 4	34. 240 x 4	50. 838 ÷ 4
3. 53 x 4	19. 624 ÷ 4	35. 345 x 4	51. 562 ÷ 4
4. 75 x 4	20. 740 ÷ 4	36. 275 x 4	52. 786 ÷ 4
5. 4 x 19	21. 340 ÷ 4	37. 4 x 185	53. 382 ÷ 4
6. 4 x 38	22. 568 ÷ 4	38. 4 x 460	54. 926 ÷ 4
7. 27 x 4	23. 520 ÷ 4	39. 280 x 4	55. 462 ÷ 4
8. 95 x 4	24. 920 ÷ 4	40. 475 x 4	56. 470 ÷ 4
9. 4 x 34	25. 680 ÷ 4	41. 4 x 630	57. 286 ÷ 4
10. 56 x 4	26. 660 ÷ 4	42. 370 x 4	58. 654 ÷ 4
11. 4 x 44	27. 344 ÷ 4	43. 4 x 235	59. 814 ÷ 4
12. 29 x 4	28. 760 ÷ 4	44. 531 x 4	60. 542 ÷ 4
13. 17 x 4	29. 588 ÷ 4	45. 175 x 4	61. 746 ÷ 4
14. 4 x 85	30. 984 ÷ 4	46. 4 x 914	62. 490 ÷ 4
15. 47 x 4	31. 388 ÷ 4	47. 724 x 4	63. 366 ÷ 4
16. 39 x 4	32. 676 ÷ 4	48. 617 x 4	64. 722 ÷ 4

Mental Math 16

Square using the ones-digit-5 method.

1. 35^2	17. 125^2
2. 75^2	18. 495^2
3. 45^2	19. 905^2
4. 85^2	20. 395^2
5. 95^2	21. 505^2
6. 25^2	22. 805^2
7. 55^2	23. 595^2
8. 15^2	24. 165^2
9. 65^2	25. 135^2
10. 115^2	26. 895^2
11. 195^2	27. 145^2
12. 205^2	28. 255^2
13. 105^2	29. 155^2
14. 305^2	30. 1005^2
15. 405^2	31. 995^2
16. 605^2	32. 705^2

Mixed Numbers 1

Convert from mixed number to fraction by mental math.		Convert from fraction to mixed number by mental math.	
1. $12\frac{13}{15}$	17. $49\frac{6}{11}$	33. $\frac{94}{15}$	49. $\frac{97}{4}$
2. $11\frac{12}{55}$	18. $5\frac{23}{32}$	34. $\frac{101}{12}$	50. $\frac{186}{19}$
3. $10\frac{16}{45}$	19. $25\frac{13}{25}$	35. $\frac{173}{13}$	51. $\frac{437}{25}$
4. $13\frac{7}{13}$	20. $17\frac{12}{17}$	36. $\frac{152}{3}$	52. $\frac{199}{18}$
5. $6\frac{10}{19}$	21. $30\frac{21}{23}$	37. $\frac{245}{6}$	53. $\frac{283}{35}$
6. $19\frac{17}{20}$	22. $12\frac{11}{12}$	38. $\frac{97}{24}$	54. $\frac{89}{22}$
7. $14\frac{9}{14}$	23. $26\frac{4}{11}$	39. $\frac{168}{11}$	55. $\frac{195}{19}$
8. $11\frac{21}{34}$	24. $11\frac{11}{53}$	40. $\frac{599}{9}$	56. $\frac{169}{11}$
9. $18\frac{17}{30}$	25. $43\frac{7}{11}$	41. $\frac{635}{7}$	57. $\frac{319}{20}$
10. $15\frac{8}{15}$	26. $40\frac{19}{20}$	42. $\frac{107}{26}$	58. $\frac{151}{35}$
11. $38\frac{8}{11}$	27. $20\frac{13}{37}$	43. $\frac{113}{14}$	59. $\frac{207}{40}$
12. $13\frac{37}{50}$	28. $25\frac{15}{31}$	44. $\frac{107}{21}$	60. $\frac{333}{31}$
13. $16\frac{15}{16}$	29. $22\frac{8}{25}$	45. $\frac{137}{17}$	61. $\frac{60}{19}$
14. $12\frac{5}{21}$	30. $14\frac{9}{10}$	46. $\frac{373}{75}$	62. $\frac{229}{55}$
15. $19\frac{9}{19}$	31. $10\frac{42}{43}$	47. $\frac{308}{51}$	63. $\frac{779}{10}$
16. $32\frac{15}{32}$	32. $15\frac{7}{12}$	48. $\frac{55}{16}$	64. $\frac{817}{9}$

Mixed Numbers 2

Multiply by converting to fractions.	*Divide by converting to fractions.*
1. $3\frac{1}{2} \times 5\frac{1}{7}$	17. $7\frac{2}{5} \div 4\frac{5}{8}$
2. $6\frac{2}{5} \times 4\frac{3}{8}$	18. $3\frac{7}{9} \div 5\frac{2}{3}$
3. $2\frac{7}{9} \times 4\frac{1}{5}$	19. $4\frac{5}{6} \div 6\frac{4}{9}$
4. $7\frac{1}{7} \times 4\frac{1}{5}$	20. $6\frac{1}{9} \div 3\frac{2}{3}$
5. $6\frac{3}{4} \times 7\frac{1}{9}$	21. $5\frac{3}{7} \div 4\frac{3}{4}$
6. $3\frac{4}{7} \times 8\frac{2}{5}$	22. $5\frac{4}{7} \div 6\frac{1}{2}$
7. $4\frac{3}{8} \times 2\frac{2}{5}$	23. $8\frac{2}{5} \div 1\frac{2}{5}$
8. $4\frac{2}{7} \times 5\frac{4}{9}$	24. $5\frac{2}{3} \div 9\frac{4}{9}$
9. $2\frac{4}{5} \times 8\frac{3}{4}$	25. $11\frac{3}{5} \div 7\frac{1}{4}$
10. $9\frac{4}{5} \times 3\frac{4}{7}$	26. $12\frac{1}{2} \div 5\frac{5}{6}$
11. $11\frac{2}{3} \times 4\frac{5}{7}$	27. $10\frac{5}{8} \div 3\frac{7}{9}$
12. $12\frac{1}{2} \times 3\frac{3}{5}$	28. $9\frac{3}{4} \div 3\frac{1}{4}$
13. $13\frac{1}{3} \times 9\frac{3}{4}$	29. $2\frac{3}{5} \div 3\frac{9}{10}$
14. $6\frac{2}{5} \times 3\frac{1}{8}$	30. $15\frac{5}{6} \div 5\frac{3}{7}$
15. $2\frac{2}{5} \times 4\frac{4}{9}$	31. $8\frac{2}{7} \div 2\frac{1}{14}$
16. $5\frac{1}{2} \times 5\frac{1}{11}$	32. $8\frac{2}{3} \div 5\frac{1}{5}$

Mixed Numbers 3

Add and subtract as indicated.

1.
$$3\frac{18}{25} + 2\frac{9}{10} - 1\frac{1}{5}$$

8.
$$9\frac{7}{9} - 1\frac{3}{10} - 2\frac{1}{6}$$

2.
$$6\frac{4}{5} + 3\frac{13}{20} - 4\frac{3}{8}$$

9.
$$9\frac{4}{5} - 3\frac{3}{10} + 4\frac{3}{16}$$

3.
$$5\frac{26}{45} + 8\frac{7}{18} - 2\frac{4}{15}$$

10.
$$4\frac{11}{15} + 2\frac{5}{12} - 1\frac{7}{30}$$

4.
$$3\frac{11}{15} + 9\frac{5}{24} - 2\frac{17}{40}$$

11.
$$1\frac{3}{7} - 4\frac{4}{21} + 8\frac{7}{9}$$

5.
$$9\frac{17}{35} + 2\frac{13}{14} - 3\frac{31}{70}$$

12.
$$8\frac{15}{33} + 5\frac{17}{22} - 4\frac{7}{11}$$

6.
$$3\frac{7}{12} + 2\frac{11}{28} + 4\frac{5}{21}$$

13.
$$5\frac{9}{25} + 1\frac{7}{15} + 2\frac{1}{5}$$

7.
$$5\frac{19}{36} + 1\frac{5}{18} + 2\frac{7}{20}$$

14.
$$5\frac{11}{33} + 1\frac{10}{11} - 2\frac{2}{9}$$

Mixed Numbers 4

Operate using DPMA and simplify.

1. $9\frac{1}{2}$ x $6\frac{2}{3}$	9. $15\frac{2}{3}$ x $6\frac{2}{5}$	17. $10\frac{5}{6}$ x $9\frac{3}{5}$
2. $5\frac{3}{4}$ x $8\frac{3}{5}$	10. $45\frac{5}{6}$ x $2\frac{8}{15}$	18. $16\frac{4}{5}$ x $5\frac{5}{8}$
3. $14\frac{1}{2}$ x $2\frac{3}{7}$	11. $36\frac{3}{4}$ x $2\frac{7}{12}$	19. $18\frac{1}{4}$ x $4\frac{8}{9}$
4. $16\frac{3}{4}$ x $4\frac{3}{8}$	12. $22\frac{1}{6}$ x $3\frac{6}{11}$	20. $21\frac{7}{8}$ x $4\frac{4}{7}$
5. $12\frac{1}{3}$ x $6\frac{1}{4}$	13. $39\frac{3}{5}$ x $2\frac{5}{13}$	21. $44\frac{2}{5}$ x $2\frac{5}{22}$
6. $10\frac{6}{7}$ x $7\frac{4}{5}$	14. $18\frac{4}{7}$ x $4\frac{7}{18}$	22. $24\frac{2}{7}$ x $3\frac{7}{12}$
7. $11\frac{5}{6}$ x $6\frac{2}{11}$	15. $15\frac{1}{4}$ x $5\frac{2}{3}$	23. $12\frac{4}{5}$ x $6\frac{3}{4}$
8. $12\frac{1}{4}$ x $4\frac{5}{12}$	16. $35\frac{1}{3}$ x $2\frac{6}{7}$	24. $30\frac{3}{4}$ x $8\frac{4}{15}$

Modulo Arithmetic 1

Complete each congruence.

1. 15 ≡ _____ (MOD 8)

2. 21 ≡ _____ (MOD 4)

3. 12 ≡ _____ (MOD 7)

4. 63 ≡ _____ (MOD 5)

5. 29 ≡ _____ (MOD 10)

6. 22 ≡ _____ (MOD 9)

7. 19 ≡ _____ (MOD 8)

8. 22 ≡ _____ (MOD 11)

9. 42 ≡ _____ (MOD 4)

10. 81 ≡ _____ (MOD 9)

11. 52 ≡ _____ (MOD 5)

12. 40 ≡ _____ (MOD 9)

13. 23 ≡ _____ (MOD 7)

14. 24 ≡ _____ (MOD 6)

15. 62 ≡ _____ (MOD 9)

16. 40 ≡ _____ (MOD 13)

17. 36 ≡ _____ (MOD 11)

18. 54 ≡ _____ (MOD 13)

19. 35 ≡ _____ (MOD 3)

20. 39 ≡ _____ (MOD 7)

21. 86 ≡ _____ (MOD 8)

22. 73 ≡ _____ (MOD 9)

23. 81 ≡ _____ (MOD 5)

24. 63 ≡ _____ (MOD 4)

25. 70 ≡ _____ (MOD 20)

26. 43 ≡ _____ (MOD 13)

27. 64 ≡ _____ (MOD 6)

28. 27 ≡ _____ (MOD 12)

29. 29 ≡ _____ (MOD 15)

30. 92 ≡ _____ (MOD 10)

31. 68 ≡ _____ (MOD 3)

32. 28 ≡ _____ (MOD 9)

33. 47 ≡ _____ (MOD 4)

34. 50 ≡ _____ (MOD 8)

35. 52 ≡ _____ (MOD 13)

36. 57 ≡ _____ (MOD 3)

37. 66 ≡ _____ (MOD 7)

38. 35 ≡ _____ (MOD 8)

39. 47 ≡ _____ (MOD 15)

40. 73 ≡ _____ (MOD 3)

41. 80 ≡ _____ (MOD 9)

42. 55 ≡ _____ (MOD 6)

43. 50 ≡ _____ (MOD 16)

44. 96 ≡ _____ (MOD 5)

45. 73 ≡ _____ (MOD 10)

46. 30 ≡ _____ (MOD 14)

47. 40 ≡ _____ (MOD 12)

48. 79 ≡ _____ (MOD 11)

Modulo Arithmetic 2

Find the opposite.	Find the reciprocal.	Find the square root.
1. $-5 \equiv$ ____ (MOD 8)	17. of 3 MOD 7 ____	33. $\sqrt{5} \equiv$ ____ (MOD 10)
2. $-4 \equiv$ ____ (MOD 5)	18. of 2 MOD 5 ____	34. $\sqrt{1} \equiv$ ____ (MOD 8)
3. $-7 \equiv$ ____ (MOD 12)	19. of 4 MOD 11 ____	35. $\sqrt{5} \equiv$ ____ (MOD 6)
4. $-5 \equiv$ ____ (MOD 7)	20. of 5 MOD 13 ____	36. $\sqrt{1} \equiv$ ____ (MOD 4)
5. $-1 \equiv$ ____ (MOD 8)	21. of 4 MOD 19 ____	37. $\sqrt{5} \equiv$ ____ (MOD 11)
6. $-7 \equiv$ ____ (MOD 14)	22. of 5 MOD 8 ____	38. $\sqrt{4} \equiv$ ____ (MOD 6)
7. $-3 \equiv$ ____ (MOD 9)	23. of 3 MOD 6 ____	39. $\sqrt{9} \equiv$ ____ (MOD 10)
8. $-8 \equiv$ ____ (MOD 10)	24. of 3 MOD 10 ____	40. $\sqrt{6} \equiv$ ____ (MOD 8)
9. $-6 \equiv$ ____ (MOD 11)	25. of 7 MOD 12 ____	41. $\sqrt{1} \equiv$ ____ (MOD 9)
10. $-5 \equiv$ ____ (MOD 6)	26. of 6 MOD 7 ____	42. $\sqrt{5} \equiv$ ____ (MOD 7)
11. $-9 \equiv$ ____ (MOD 12)	27. of 9 MOD 11 ____	43. $\sqrt{0} \equiv$ ____ (MOD 8)
12. $-6 \equiv$ ____ (MOD 13)	28. of 9 MOD 10 ____	44. $\sqrt{5} \equiv$ ____ (MOD 11)
13. $-4 \equiv$ ____ (MOD 7)	29. of 3 MOD 13 ____	45. $\sqrt{6} \equiv$ ____ (MOD 10)
14. $-9 \equiv$ ____ (MOD 16)	30. of 13 MOD 19 ____	46. $\sqrt{2} \equiv$ ____ (MOD 7)
15. $-2 \equiv$ ____ (MOD 15)	31. of 4 MOD 5 ____	47. $\sqrt{1} \equiv$ ____ (MOD 6)
16. $-0 \equiv$ ____ (MOD 8)	32. of 4 MOD 9 ____	48. $\sqrt{3} \equiv$ ____ (MOD 6)

Modulo Arithmetic 3

Answer YES or NO as to whether the MOD has the specified property. See Abbreviations.

	MOD 2	MOD 3	MOD 4	MOD 5	MOD 6	MOD 7	MOD 8	MOD 9	MOD 10	MOD 11	MOD 12
1. CíPA											
2. APA											
3. IdPA											
4. InPA											
5. CPA											
6. CíPM											
7. APM											
8. IdPM											
9. InPM											
10. CPM											
11. DPMA											
12. ZPM											
13. RPC											
14. SPC											
15. TPC											
16. APC											
17. MPC											

Modulo Arithmetic 4

Find the specified numbers in the mods.

1. $\frac{1}{2} \equiv$ ____ (MOD 5)	17. $\frac{2}{3} \equiv$ ____ (MOD 19)	33. $\sqrt{-2} \equiv$ ____ (MOD 7)
2. $\frac{2}{3} \equiv$ ____ (MOD 5)	18. $\frac{2}{3} \equiv$ ____ (MOD 17)	34. $\sqrt{-4} \equiv$ ____ (MOD 5)
3. $\frac{1}{2} \equiv$ ____ (MOD 7)	19. $\frac{4}{9} \equiv$ ____ (MOD 11)	35. $\sqrt{-5} \equiv$ ____ (MOD 7)
4. $\frac{2}{3} \equiv$ ____ (MOD 7)	20. $\frac{3}{7} \equiv$ ____ (MOD 17)	36. $\sqrt{-3} \equiv$ ____ (MOD 7)
5. $\frac{1}{2} \equiv$ ____ (MOD 11)	21. $\frac{5}{6} \equiv$ ____ (MOD 19)	37. $\sqrt{-7} \equiv$ ____ (MOD 11)
6. $\frac{2}{3} \equiv$ ____ (MOD 11)	22. $\frac{3}{8} \equiv$ ____ (MOD 11)	38. $\sqrt{-5} \equiv$ ____ (MOD 10)
7. $\frac{1}{2} \equiv$ ____ (MOD 13)	23. $\frac{3}{8} \equiv$ ____ (MOD 13)	39. $\sqrt{-9} \equiv$ ____ (MOD 13)
8. $\frac{2}{3} \equiv$ ____ (MOD 13)	24. $\frac{4}{9} \equiv$ ____ (MOD 13)	40. $\sqrt{-2} \equiv$ ____ (MOD 6)
9. $\frac{3}{4} \equiv$ ____ (MOD 11)	25. $\sqrt{-1} \equiv$ ____ (MOD 5)	41. $\sqrt{-1} \equiv$ ____ (MOD 11)
10. $\frac{3}{4} \equiv$ ____ (MOD 7)	26. $\sqrt{-1} \equiv$ ____ (MOD 7)	42. $\sqrt{-1} \equiv$ ____ (MOD 13)
11. $\frac{2}{5} \equiv$ ____ (MOD 11)	27. $\sqrt{-1} \equiv$ ____ (MOD 17)	43. $\sqrt{-4} \equiv$ ____ (MOD 8)
12. $\frac{3}{5} \equiv$ ____ (MOD 13)	28. $\sqrt{-4} \equiv$ ____ (MOD 13)	44. $\sqrt{-1} \equiv$ ____ (MOD 10)
13. $\frac{2}{5} \equiv$ ____ (MOD 7)	29. $\sqrt{-6} \equiv$ ____ (MOD 7)	45. $\sqrt{-4} \equiv$ ____ (MOD 7)
14. $\frac{5}{6} \equiv$ ____ (MOD 11)	30. $\sqrt{-2} \equiv$ ____ (MOD 17)	46. $\sqrt{-2} \equiv$ ____ (MOD 11)
15. $\frac{5}{6} \equiv$ ____ (MOD 13)	31. $\sqrt{-13} \equiv$ ____ (MOD 17)	47. $\sqrt{-4} \equiv$ ____ (MOD 10)
16. $\frac{4}{7} \equiv$ ____ (MOD 11)	32. $\sqrt{-8} \equiv$ ____ (MOD 11)	48. $\sqrt{-8} \equiv$ ____ (MOD 17)

Money 1

Find the number of each coin using smart trial and error.	*Answer as indicated.*
1. 17 quarters and dimes have a value of $3.05.	8. 7 coins = 58¢ How many quarters?
2. 16 quarters and nickels have a value of $2.40.	9. 8 coins = 37¢ How many dimes?
3. 29 nickels and dimes have a value of $2.05.	10. 10 coins = 69¢ How many nickels?
4. 22 quarters and dimes have a value of $3.70.	11. 9 coins = 135¢ How many dimes?
5. 40 nickels and dimes have a value of $2.75.	12. 10 coins = 37¢ How many nickels?
6. 20 quarters and nickels have a value of $3.60.	13. 10 coins = 110¢ How many nickels?
7. 28 quarters and dimes have a value of $4.30.	14. 10 coins = 100¢ How many nickels?

Money 2

Find the number of coins (at least 1 of each) for the given value and conditions.	*Find the number of different ways to make $1 using . . .*
1.　pennies, dimes, and quarters to make $1.67; fewest number of coins	8.　quarters and dimes.
2.　nickels and dimes to make $1.75; twice as many dimes as nickels	9.　quarters and nickels.
3.　dimes and quarters to make $2.50; 3 more quarters than dimes	10.　dimes and nickels.
4.　nickels and dimes to make $2.20; twice as many nickels as dimes	11.　dimes and pennies.
5.　dimes and quarters to make $1.95; 9 more dimes than quarters	12.　quarters, dimes, and nickels.
6.　pennies, nickels, dimes, and quarters to make $7.98; fewest number of coins	13.　half dollars, nickels, and pennies.
7.　nickels and quarters to make $1.60; 3 times as many nickels as quarters	14.　half dollars, quarters, and dimes.

Money 3

Answer as indicated.

Find the total number of coins given an equal number of each.

1. 14 quarters, 15 dimes, and 40 nickels is what percent of $5?

8. nickels, dimes, and quarters
value = $2.00

2. 19 quarters, 16 dimes, and 55 pennies is what percent of $10?

9. pennies, nickels, and dimes
value = $4.00

3. 21 quarters, 21 dimes, and 18 nickels is what percent of $20?

10. pennies, dimes, and quarters
value = $2.52

4. 22 quarters, 19 dimes, and 55 nickels is what percent of $50?

11. nickels, dimes, and half-dollars
value = $7.15

5. 11 quarters, 9 dimes, and 18 nickels is what percent of $5?

12. nickels, quarters, and half-dollars
value = $10.40

6. 16 quarters, 19 dimes, and 15 pennies is what percent of $10?

13. dimes, quarters, and half-dollars
value = $10.20

7. 26 quarters, 33 dimes, and 7 nickels is what percent of $20?

14. pennies, nickels, and quarters
value = $4.34

Money 4

Find the number coins using algebra.	*Find the sum of the values that may be made using one or more of the coins.*
1. A group of 57 coins, only nickels and quarters, has value $7.05. Find the number of each coin.	6. 1 quarter, 1 dime, 1 penny
2. A group of dimes and quarters, with 4 more quarters than dimes, is worth $13.60. Find the number of each.	7. 1 dime, 1 nickel, 1 penny
3. A group of 46 coins, only nickels and quarters, has value $8.70. Find the number of each coin.	8. 1 quarter, 1 nickel, 1 penny
4. A group of dimes and quarters, with 9 more dimes than quarters, is worth $8.95. Find the number of each.	9. 1 dime, 2 nickels, 1 penny
5. A group of coins, with twice as many dimes as quarters and no others, has $10.80 value. Find how many of each.	10. 2 quarters, 1 dime, 1 penny

Order of Operations 1

Operate. Show one line of work before the answer.

1. $2 + 2 \times 2 - 2 \div 2 \times 2 + 2 - 2 + 2^2$

2. $18 \div 6 \times 3 - 2^3 + (2 \times 4) + 18 \div 2$

3. $7 + 5(1 - 5) - 2 \times (3 - 8) + 6^2 \div 3$

4. $3 \div 3 \times 3 \times 3 \div 3 - 3 \times 3 \times 3 \div 3^2$

5. $9 + 8 \div 4 - 3 \times 4 \div 6 - 45 \div 3 - 1$

6. $3(12 \div 3) - 20 \div 5 + 3 - 5 \times 4 - 2$

7. $3 - 2(5 - 7)^3 - (3 - 12) + 40 \div 8$

8. $2 \times 5^2 - 30 \div 3 \times 5 + 60 \div 12 \times 5$

9. $4 - 4(3 - 8)^2 - 2 \times (7 - 8) + 5^2$

10. $3 \times (7 - 7) - 3^3 - (3 + 7) + 14 \div 7$

11. $3 \times 4^2 - 80 \div 5 \times 2 - 35 \div 7 \times 5$

12. $6 - 6 \div 3 - 6 \times 3 \div 6 - 36 \div 3 - 3$

13. $1 - (7 - 1)^2 - 1 + 7 \times (1 - 7) - 1^7$

14. $8 + 8 - 8 \times 8 \div 8 - 8 + 8 \div 8 - 8^2$

15. $11 + 9 \div 3 - 7 \times 4 \div 2 - 60 \div 5 - 2$

16. $-2(2 - 5)^3 - (7 - 12) + 52 \div 4 - 4$

52

Order of Operations 2

Operate. Show one line of work before the answer.

1. $10 \div 3 \times 6 - 8 \div 5 \times 10 - 2 \div 2^2$

9. $7 - 5(3 - 6)^2 - 2 \div (7 - 12) \times 10^2$

2. $13 \div 3 \times 9 \times 2 - (1 - 4)^3 - (4 - 8)$

10. $9 \times (7 - 8) - 4^3 - (3 + 8) \div 22 \times 2$

3. $-1 - 4 - 6 \times (3 - 8) \div 7 \times 21 - 5^2$

11. $-4 \times 5^2 - 7 \div 3 \times 33 - 7 \div 10 \times 5$

4. $4 \div 3 \times 9 \div 5 \times 15 - 9 \times 9 \div 3^2 \times 2$

12. $16 - 13 \div 7 \times 21 \div 5 \times 10 - 2 \div 5$

5. $(5 + 12 \div 5 \times 20 \times 2 - 6) \div 5 - 10$

13. $10 - (3 - 10)^2 - 1 \div 7 \times (1 - 15)$

6. $2(11 \div 5 \times 20) \div 8 - 2(5 \times 4 - 2)$

14. $6 \div 5 \times 20 \div 7 \times 28 - 6 - (5 - 5^2)$

7. $-3(6 - 8)^3 - 11 \div 3 \times 15 + 1 \div 4$

15. $5 \div 3 \times 12 - 4 \div 8 \times 7 - 9 \div 18 + 1$

8. $3 \times 4^2 - 21 \div 5 \times 15 - 7 \div 14 \times 60$

16. $-2(2 - 7)^3 - (5 - 11) \div 4 \div 24 \times 16$

MAVA Math: Enhanced Skills   Copyright © 2015 Marla Weiss

Parallelograms 1

Find the area in square units of parallelogram ABCD with the given vertices.

1. A(4, 4), B(−8, 4),
   C(−14, −4), D(−2, −4)

2. A(−3, 7), B(6, 7),
   C(−2, −5), D(−11, −5)

3. A(20, −5), B(−2, −5),
   C(−7, 6), D(15, 6)

4. A(1, 10), B(13, 10),
   C(10, −2), D(−2, −2)

5. A(0, 0), B(4, 7),
   C(13, 7), D(9, 0)

6. A(8, 5), B(−5, 5),
   C(−11, −4), D(2, −4)

7. A(0, −3), B(3, 9),
   C(17, 9), D(14, −3)

8. A(0, 0), B(−4, 0),
   C(9, 7), D(13, 7)

9. A(9, 6), B(−7, 6),
   C(−13, −7), D(3, −7)

10. A(0, −3), B(3, 9),
    C(17, 9), D(14, −3)

11. A(−3.5, 8), B(6.5, 8),
    C(−1, −8), D(−11, −8)

12. A(18.4, −1), B(2.4, −1),
    C(7, 6.5), D(23, 6.5)

13. A(2, −4), B(−8, −4),
    C(−14, −9), D(−4, −9)

14. A(13, 9), B(−7, 9),
    C(−16, −7), D(4, −7)

15. A(0.5, −3.5), B(2.5, 9.5),
    C(7.5, 9.5), D(5.5, −3.5)

16. A(1, 10.3), B(14, 10.3),
    C(11, −2.7), D(−2, −2.7)

17. A(−8, 7), B(7, 7),
    C(−1, −8), D(−16, −8)

18. A(20, −3), B(2, −3),
    C(4, 8), D(22, 8)

19. A(2, −2), B(−10, −2),
    C(−15, −9), D(−3, −9)

20. A(2, 11), B(−7, 11),
    C(−14, −6), D(−5, −6)

Parallelograms 2

Find the four angles of parallelogram ABCD.

1. $m\angle A = 8x - 5$
   $m\angle B = 2x + 23$

2. $m\angle A = 2x + 3$
   $m\angle C = -x + 21$

3. $m\angle A = 7x - 4$
   $m\angle B = 5x + 10$

4. $m\angle A = 6x + 11$
   $m\angle C = -2x + 35$

5. $m\angle A = 6x - 5$
   $m\angle B = 9x + 20$

6. $m\angle A = 2x + 3y$
   $m\angle B = -x + 4y$
   $m\angle C = 6x + y$

7. $m\angle A = 4x + y$
   $m\angle B = -2x + 7y$
   $m\angle C = 2x + 3y$

8. The measure of one angle of the parallelogram is 5 more than 4 times another.

9. The sum of 3 of the angles of the parallelogram is 40 more than 3 times the fourth.

10. The measure of one angle of the parallelogram is 4 less than 7 times another.

11. The sum of 3 of the angles of the parallelogram is 60 less than 3 times the fourth.

12. The measure of one angle of the parallelogram is 16 more than 3 times another.

13. The measure of one angle of the parallelogram is 2 less than 6 times another.

14. The sum of 3 of the angles of the parallelogram is 78 more than twice the fourth.

Parallelograms 3

Find the perimeter of parallelogram ABCD in units. Angles given are for parallelograms. NTS

1. AB = 9x – 4
    BC = 11
    AD = 10x + 1

2. AB = 2x + 3
    BC = 10x + 1
    CD = 18x – 5

3. BA = 5x – 2
    CD = 9x – 6
    AD = 4x + 1

4. BA = 5x + 3
    BC = 17
    AD = 10x – 3

5. BA = 6x – 5
    AD = 3x + 5
    DC = 25

6. BC = 6x + 4
    DC = 4x + 1
    DA = 8x – 3

7. AB = 8x + 4
    CD = 10x – 1
    BC = 6x + 7

8. BA = 2x + 18
    BC = 26
    DA = 7x + 5

9. BC = 11x – 8
    DC = 4x + 1
    DA = 5x + 16

10. one angle 30º
    BE = 6
    BC = 12

11. one angle 45º
    BE = 10
    AD = 30

12. BE = 12
    AE = 9
    DE = 11

13. one angle 45º
    BE = 14
    AD = 27

14. one angle 60º
    BE = 12
    BC = 15

15. BE = 24
    AE = 10
    DE = 13

16. perimeter of ΔABD = 33
    BF = 5

17. perimeter of ΔCAD = 44
    AF = 9

18. altitude from B to $\overline{CD}$ = 16
    BC = 28
    BE = 20

Parallelograms 4

Answer as indicated. Linear measurements are in units.

---

1. Randomly select two angles of a parallelogram. Find the probability that they are congruent.

6. The perimeter of parallelogram ABCD is 320. AB:BC = 11:5. Find AB.

---

2. Randomly select two angles of a parallelogram. Find the probability that they are congruent or supplementary.

7. The perimeter of parallelogram ABCD is 540. BC:CD = 8:7. Find CD.

---

3. $\overline{AE}$ and $\overline{DE}$ are angle bisectors of parallelogram ABCD. Find m$\angle$AED.

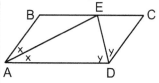

8. The perimeter of parallelogram ABCD = 70. The altitude from B to $\overline{CD}$ = 16. Find the altitude from B to $\overline{AD}$.

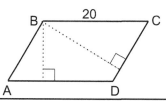

---

4. The perimeter of parallelogram ABCD is 88. The perimeter of FECD is 56. EC = AF. Find EF.

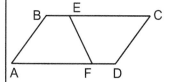

9. The perimeter of parallelogram ABCD is 64. The perimeter of ABEF is 47. BE = FD. Find EF.

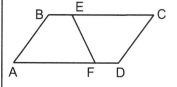

---

5. A parallelogram is inscribed in a circle. What must be true about the parallelogram?

10. The perimeter of parallelogram ABCD is 160. AB:BC = 3:2. Find BC.

Percents 1

Find the percent by mental math.

1. 10% of 50	17. 3% of 800	33. 12.5% of 16
2. 10% of 660	18. 27% of 1000	34. 75% of 84
3. 20% of 200	19. 25% of 500	35. 11% of 900
4. 30% of 20	20. 75% of 60	36. 4% of 700
5. 5% of 300	21. 40% of 25	37. 30% of 160
6. 15% of 500	22. 30% of 5000	38. 66 2/3% of 33
7. 25% of 48	23. 37.5% of 80	39. 80% of 170
8. 50% of 102	24. 100% of 561	40. 15% of 80
9. 90% of 450	25. 15% of 2000	41. 50% of 846
10. 12.5% of 48	26. 20% of 570	42. 25% of 440
11. 33 1/3% of 99	27. 70% of 500	43. 12% of 600
12. 13% of 100	28. 50% of 64	44. 37.5% of 888
13. 1% of 300	29. 15% of 880	45. 75% of 32
14. 2% of 150	30. 10% of 1210	46. 12.5% of 800
15. 90% of 900	31. 200% of 45	47. 30% of 120
16. 50% of 350	32. 33 1/3% of 990	48. 25% of 1000

Percents 2

Complete the chart of equivalent values. Simplify fractions.

PERCENT	DECIMAL	FRACTION	PERCENT	DECIMAL	FRACTION
1. 12%			17. 52%		
2.	.34		18.		$\frac{18}{25}$
3.		$\frac{16}{25}$	19.	.39	
4. 32%			20.		$\frac{13}{20}$
5.	.98		21.	.28	
6.	2.4		22. 15%		
7.		$\frac{11}{20}$	23. 435%		
8. 16%			24.		$\frac{27}{50}$
9.	.45		25. 106%		
10.	.03		26.	.74	
11.		$\frac{33}{50}$	27. 8%		
12. 78%			28.		$\frac{21}{25}$
13. 30%			29. 24%		
14.		$\frac{17}{20}$	30.	1.05	
15. 245%			31.		$\frac{19}{20}$
16.		$\frac{14}{25}$	32.	1.44	

Percents 3

Answer each "forward" percent word problem using written mental math.

1. Find the price of a
   $640 keyboard sold
   at a 25% discount
   with 5% sales tax.

2. Find the percent discount
   on the total purchase
   of a $30 book at a 20%
   discount and a $20 book
   at a 30% discount.

3. Find the price of a
   $350 suit sold at a
   30% discount with
   5% sales tax.

4. With 4% sales tax,
   find the total cost of 8
   items at $13.25 each.

5. With 3% sales tax,
   find the total cost of
   a $9.75 item and a
   $13.25 item.

6. With 6% sales tax,
   find the total cost of
   six items at $11.25
   each.

7. Find the price of an
   $800 item sold at a
   37.5% discount with
   4% sales tax.

8. Find the price of a
   $450 chair sold at a
   20% discount with
   5% sales tax.

9. Find the total cost if
   each of 20 people has
   a bill of $12.50 plus
   5% tax and 15% tip.

10. Tim earns $175 a week
    base plus 15% commission
    on sales. Find his weekly
    earnings when his sales for
    4 weeks are $2640.

11. Find the price of a
    $900 item sold at a
    20% discount with
    5% sales tax.

12. Lily earns $220 a week
    base salary plus 12%
    commission. Find her
    total earnings in a week
    when her sales are $850.

13. If a population of
    64,000 decreases by
    25% each year, find
    the population after 2
    years.

14. Find the percent
    discount when an
    item that retails for
    $4000 is on sale for
    $3480.

15. The value of a multi-media
    system depreciates by 20%
    a year. If it was purchased
    2 years ago for $8000, find
    its current value.

16. Find the price of a
    $240 item sold at a
    35% discount with
    10% sales tax.

Percents 4

Complete the chart of equivalent values. Simplify fractions.

	PERCENT	DECIMAL	FRACTION		PERCENT	DECIMAL	FRACTION
1.	12.5%			17.	.75%		
2.		.008		18.			$\dfrac{1}{400}$
3.			$\dfrac{3}{8}$	19.		.036	
4.	3.25%			20.			$\dfrac{7}{8}$
5.		.005		21.		.075	
6.		6.5		22.	.4%		
7.			$\dfrac{13}{2000}$	23.	1.75%		
8.	.6%			24.			$\dfrac{27}{500}$
9.		.045		25.	.35%		
10.		.0875		26.		.3125	
11.			$\dfrac{33}{400}$	27.	3.75%		
12.	2.5%			28.			$\dfrac{11}{400}$
13.	.1%			29.	5.5%		
14.			$\dfrac{17}{200}$	30.		.0625	
15.	62.5%			31.			$\dfrac{3}{200}$
16.			$\dfrac{1}{500}$	32.		.022	

Percents 5

Answer as indicated by forming "is/of," following the English into an equation, writing a proportion, or forming a 2-column list.

1. 16 is what percent of 20?	10. 75% of what number is 120?
2. 20 is what percent of 16?	11. 75 is what percent of 120?
3. 16 is 20% of what number?	12. What percent of 40 is 84?
4. 20 is 16% of what number?	13. 84 is 40% of what number?
5. 2.4% of what number is 12?	14. 36 is 30% of what number?
6. 2.4 is what percent of 12?	15. What percent of 150 is 48?
7. 2.4 is 12% of what number?	16. 52 is 4% of what number?
8. 35% of what number is 56?	17. 12 is 150% of what number?
9. 35 is what percent of 56?	18. What percent of 225 is 90?

Percents 6

Find the percent increase (I) or decrease (D) using change/original.

1. from 25 to 28	12. from 25 to 80	23. from 40 to 35
2. from 480 to 720	13. from 9 to 17	24. from 10 to 50
3. from 20 to 11	14. from 82 to 205	25. from 50 to 42
4. from 45 to 50	15. from 55 to 75	26. from 32 to 40
5. from 1000 to 20	16. from 50 to 80	27. from 11 to 13
6. from 560 to 420	17. from 18 to 12	28. from 19 to 57
7. from 360 to 480	18. from 60 to 160	29. from 20 to 50
8. from 99 to 66	19. from 75 to 100	30. from 60 to 54
9. from 140 to 560	20. from 320 to 200	31. from 12 to 20
10. from 40 to 15	21. from 480 to 72	32. from 20 to 12
11. from 72 to 36	22. from 500 to 50	33. from 55 to 60

Percents 7

Represent the value as a fraction.

1. 5% increase	20. 40% decrease
2. 5% decrease	21. 45% increase
3. 10% increase	22. 45% decrease
4. 10% decrease	23. 50% increase
5. 12.5% increase	24. 50% decrease
6. 12.5% decrease	25. 60% increase
7. 15% increase	26. 60% decrease
8. 15% decrease	27. 62.5% increase
9. 20% increase	28. 62.5% decrease
10. 20% decrease	29. 66 2/3% increase
11. 25% increase	30. 66 2/3% decrease
12. 25% decrease	31. 70% increase
13. 30% increase	32. 70% decrease
14. 30% decrease	33. 75% increase
15. 33 1/3% increase	34. 75% decrease
16. 33 1/3% decrease	35. 80% increase
17. 37.5% increase	36. 80% decrease
18. 37.5% decrease	37. 90% increase
19. 40% increase	38. 90% decrease

Percents 8

Answer the "backward" percent word problem. Use a calculator after writing the set-up.

1. Find the original price of an item that costs $12.60 after a 16% discount.	9. The price of a chair with a 15% discount was $187. What was the original price?
2. A book was bought on sale for $40.95 at 9% off. What was the regular price?	10. $270.40 is in a bank paying 4% simple annual interest. Find the value of the account 1 year ago assuming no other activity.
3. A game was purchased for $32.19 at 13% off. What was the original price?	11. A shirt was bought for $15.90 including the 6% tax. What was the price without the tax?
4. A shirt cost $18 after a 20% discount. What was the original price?	12. Find the original price of an item that costs $63 after a 16% discount.
5. A coat cost $132 after a 12% discount. What was the original price?	13. The price of a stereo set with a 20% discount was $924. Find the original price.
6. The price of a sofa with 5% tax was $924. What was the price without tax?	14. A boat costs $13,250 with the 6% sales tax. What was the price prior to the tax?
7. A television was bought for $507 at 22% off the regular price. Find the regular price.	15. Jake spent $380 for an item at 24% off the regular price. What was the regular price?
8. An item cost $48.15 including the 7% sales tax. What was the price without the tax?	16. $705.25 is invested paying 8.5% simple annual interest. Find the value of the account 1 year ago assuming no other activity.

Percents 9

Find the final percent increase (I) or decrease (D) after successive changes.

	FRACTION METHOD	100 METHOD	ANSWER
1. 50% I, 50% D			
2. 30% I, 30% D			
3. 25% I, 20% I			
4. 20% D, 15% D			
5. 25% D, 20% D			
6. 50% I, 33 1/3% D			
7. 50% I, 33 1/3% I			
8. 70% D, 33 1/3% I			
9. 75% D, 20% D			
10. 50% D, 50% D			
11. 90% D, 40% I			
12. 100% I, 50% I			
13. 20% I, 20% D			
14. 40% D, 33 1/3% I			
15. 20% I, 25% D, 30% D			
16. 25% D, 66 2/3% I, 20% D			
17. 60% I, 80% D, 50% I			
18. 20% D, 37.5% I, 20% I			
19. 20% D, 20% I, 50% D			

Percents 10

Answer as indicated.

1.	A price increases by 10% and then again by 10% to $6.05. What was the original price?	7.	A price decreases by 30% and then increases by 40%. Find the ratio of the last price to the original.
2.	A salary is reduced by 20%. Find the percent increase to return it to the original amount. 2nd method	8.	A price increases by 30% and then by 40% to $662.48. What was the original price?
3.	A price increases by 25% and then decreases by 40%. Find the ratio of the last price to the original.	9.	A price increases by 20% and then again by 20% to $100.80. What was the original price?
4.	A price increases by 40% and then decreases by 25%. Find the ratio of the last price to the original.	10.	A salary is reduced by 25%. Find the percent increase to return it to the original amount.
5.	A price increases by 20% and then again by 20% to $86.40. What was the original price?	11.	A number is reduced by 60%. Find the percent increase to return it to the original value.
6.	A price increases by 75% and then decreases by 60%. Find the ratio of the last price to the original.	12.	A salary is reduced by 50%. Find the percent increase to return it to the original amount.

Percents 11

Find how much more must be added to the given amount to make 100%.	Find a percent of a percent.
1. 4800 is 60%.	14. 35% is what percent of 80%?
2. 135 is 30%.	15. 25% is what percent of 60%?
3. 640 is 40%.	16. 30% is what percent of 80%?
4. 21 is 5%.	17. 30% is what percent of 90%?
5. 6300 is 70%.	18. 50% is what percent of 75%?
6. 3434 is 85%.	19. 60% is what percent of 72%?
7. 609 is 15%.	20. 3% is what percent of 48%?
8. 763 is 35%.	21. 30% is what percent of 96%?
9. 459 is 45%.	22. 15% is what percent of 90%?
10. 7742 is 70%.	23. 49% is what percent of 84%?
11. 303 is 15%.	24. 60% is what percent of 96%?
12. 2639 is 65%.	25. 44% is what percent of 99%?
13. 352 is 55%.	26. 12% is what percent of 64%?

Percents 12

Answer as indicated.

1. 7 out of 19 marbles are pink. How many pink marbles must be added to make the number of pinks 52%?	8. If x is 40% of y, what % of 8y is 7x?
2. 20 out of 31 marbles are blue. How many blue marbles must be added to make the number of blues 78%?	9. If x is 30% of y, what % of 3y is 5x?
3. 41 out of 63 marbles are red. How many red marbles must be removed to make the number of reds 56%?	10. If x is 60% of y, what % of 3y is 4x?
4. 9 out of 18 buttons are white. How many white buttons must be added to make the number of whites 64%?	11. If x is 140% of y, 2x is what % of 7y?
5. 10 out of 17 peppers are green. How many green peppers must be added to make the number of greens 72%?	12. If x is 90% of y, what % of 9y is 2x?
6. 37 out of 64 marbles are red. How many red marbles must be removed to make the number of reds 46%?	13. If x is 80% of y, 4x is what % of 8y?
7. 20 out of 31 marbles are blue. How many blue marbles must be removed to make the number of blues 45%?	14. If x is 70% of y, 4x is what % of 8y?

Percents 13

Evaluate using fraction simplification.

	Find the percent change in the area of a:

1. 10% of 20% of 40% of 250

10. rectangle if each side is increased by 20%.

2. 25% of 20% of 75% of 64,000

11. rectangle if 1 side is increased by 50% and 1 side is increased by 100%.

3. 10% of 20% of 35% of 50,000

12. square if each side is increased by 50%.

4. 30% of 40% of 80% of 7500

13. rectangle if 1 side is decreased by 20% and 1 side is increased by 50%.

5. 25% of 10% of 30% of 60,000

14. rectangle if each side is decreased by 40%.

6. 15% of 60% of 12.5% of 84,000

15. rectangle if each side is increased by 40%.

7. 20% of 12% of 87.5% of 9000

16. square if 1 side is increased by 40% and 1 side is decreased by 40%.

8. 45% of 28% of 37.5% of 80,000

17. rectangle if each side is increased by 30%.

9. 16% of 65% of 62.5% of 100,000

18. rectangle if each side is increased by 10%.

Percents 14

Answer as indicated.

Answer as indicated for the solid figures.

1. If 15% of a number is subtracted from the number, the result is 68. Find the number.

6. A square pyramid has its altitude tripled and each edge of its base halved. Find the percent change in its volume.

2. If 45% of a number is added to the number, the result is 580. Find the number.

7. A square pyramid has its altitude halved and each edge of its base tripled. Find the percent change in its volume.

3. If 36% of a number is subtracted from the number, the result is 88. Find the number.

8. A rectangular pyramid with L=2W has its altitude halved and each edge of its base doubled. Find the percent change in its volume.

4. If 400 critters are 13 1/3% of the critter population, how many are 100%?

9. A rectangular pyramid with L=4W has its altitude doubled and each edge of its base halved. Find the percent change in its volume.

5. If 252 critters are 11 2/3% of the critter population, how many are 100%?

10. A solid has a pentagonal base of area 72, total altitude 20, and pyramidal top with altitude 8. The volume of the top is what percent of the volume of the solid?

Perfect Squares & Cubes 1

Find the least natural number n such that the product is a perfect square.	*Find the least natural number n such that the product is a perfect cube.*
1. 98n	14. 98n
2. 160n	15. 160n
3. 540n	16. 540n
4. 343n	17. 343n
5. 48n	18. 48n
6. 360n	19. 360n
7. 500n	20. 500n
8. 108n	21. 108n
9. 504n	22. 504n
10. 144n	23. 144n
11. 180n	24. 180n
12. 250n	25. 250n
13. 320n	26. 320n

Perfect Squares & Cubes 2

Answer as indicated.

1. The sum of the first 20 positive perfect squares is 2870. Find the sum of the first 19 positive perfect squares.

2. Find the product of the 50th and 100th positive perfect squares.

3. Find the least positive integer n for which 5n is both an even integer and a perfect square.

4. Find the least perfect square divisible by the first 4 prime numbers.

5. Find the natural numbers less than 200 that have exactly 3 factors.

6. The product mn is a perfect square. Both m and n are neither prime nor square. Find the least possible sum m + n.

7. The sum of the squares of three consecutive positive integers is 677. Find the sum of the integers.

8. How many perfect squares are between 400 and 3600?

9. A perfect cube was increased by 700%. By what percent was its cube root increased?

10. The four-digit number 7AB1 is a perfect square. Find A + B.

11. Find the greatest 4-digit number that has exactly 3 factors.

12. The sum of five consecutive integers is a perfect square. The sum of six consecutive integers is a perfect cube. Find the least product of the two sums.

Perimeter 1

Calculate perimeter in units. Assume right angles and semicircles.

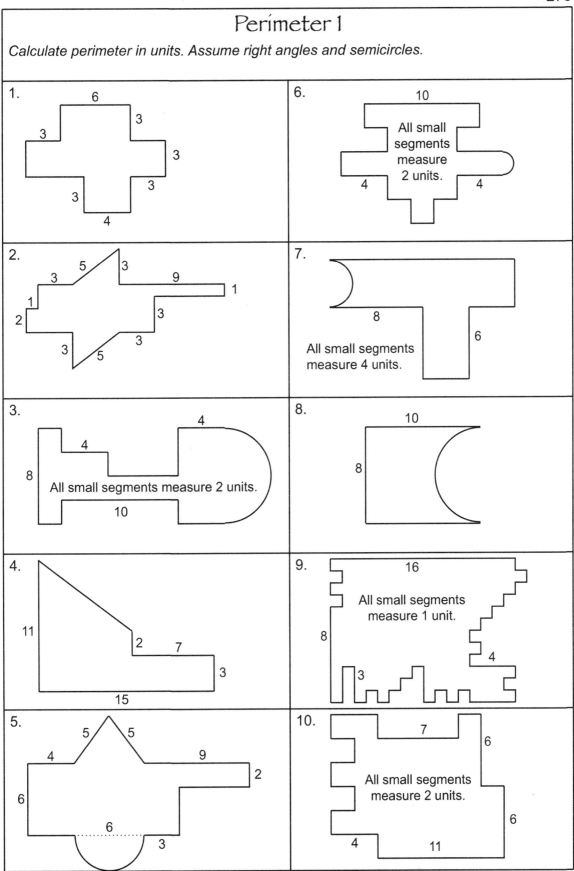

1.

6
3
3
3
3
3
4

2.

5 3
3
1
9
1
2
3
3
3
5

3.

4
8
4
All small segments measure 2 units.
10

4.

11
2
7
3
15

5.

5 5
4
9
2
6
6
3

6.

10
All small segments measure 2 units.
4
4

7.

8
6
All small segments measure 4 units.

8.

10
8

9.

16
All small segments measure 1 unit.
8
4
3

10.

7
6
All small segments measure 2 units.
6
4
11

Perimeter 2

Answer as indicated. All lengths are in units.

1. Find the perimeter of a square with side 0.25.	9. The perimeter of a rectangle is 1/5. Find the sum of two adjacent sides.
2. The perimeter of a rectangle is 1/9. Find the sum of two adjacent sides.	10. A square has area 144. Triple the length of one side and halve the other. Find the ratio of the perimeters: square to rectangle.
3. The figure with area 294 consists of 6 congruent squares. Find the perimeter.	11. A rectangle with L 6 more than its W gets a uniform border of 5 units added to all 4 sides. The new perimeter is 104. Find the original perimeter.
4. If a square has area 49, find its semi-perimeter.	12. Find the perimeter of the figure: an equilateral triangle is attached at one side of a square with area 81.
5. Find the sum of the perimeters: a square with diagonal 12 and a square with diagonal $12\sqrt{2}$.	13. A 6 by 8 rectangle has its upper right corner "punched in" at right angles to form an "L" shape. Find the perimeter of the new figure.
6. A rectangle with W 4 times its L gets a uniform border of 3 units added to all 4 sides. The new perimeter is 174. Find the original perimeter.	14. A square has area 400. Double one side and quarter the other. Find the ratio of the perimeters: rectangle to square.
7. If two equilateral triangles with side 3.25 are attached at one side, find the perimeter of the quadrilateral formed.	15. Four squares are attached to form a rectangle with shortest side 5. Find the perimeter of the rectangle.
8. Find the perimeter of the figure that is 3/4 of a square with area 100.	16. A square is divided into four congruent smaller squares. Find the perimeter of a small square if the perimeter of the large square is 2.

Perimeter 3

Answer as indicated.

1. Find the perimeter of a regular octagon if a side has length 5.125.	8. Find the perimeter of an equilateral triangle if 5/6 the perimeter is 45.
2. Find the perimeter of an equilateral triangle that is attached at one side of a square with area 81.	9. Find the perimeter of a regular pentagon if a side has length 1.2.
3. Find the perimeter of an equilateral triangle with area $16\sqrt{3}$.	10. Find the side of an equilateral triangle if a scalene triangle with sides 5, 6, and 10 has the same perimeter.
4. Find the perimeter of an equilateral triangle with area $12\sqrt{3}$.	11. Find the perimeter of an equilateral triangle if 2/3 the perimeter is 12.
5. The ratio of 2 adjacent sides of a parallelogram is 5:4. If the perimeter is 117, find the shorter side.	12. The ratio of 2 adjacent sides of a parallelogram is 7:3. If the perimeter is 170, find the longer side.
6. Find the greatest possible perimeter of a right triangle with area 10 if the legs are whole numbers.	13. Find the least possible perimeter of a right triangle with area 1800 if the legs are whole numbers.
7. Find the perimeter of a regular hexagon with area $24\sqrt{3}$.	14. Find the perimeter of a regular hexagon with area $30\sqrt{3}$.

Perimeter 4

Calculate the perimeter in units. Neighboring grid lines are one unit apart.

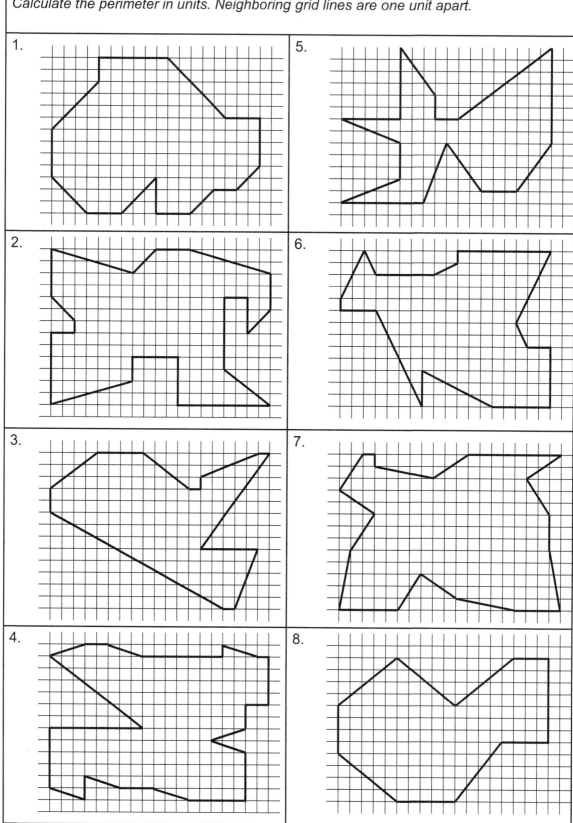

Permutations 1

Find the number of arrangements of the letters of the word.	Answer as indicated using the multiplication principle.
1. MATH	11. Find the number of arrangements of 4 letters of the word EDUCATION.
2. FRACTIONS	12. Find the number of ways to arrange 6 different books on a shelf.
3. PRIME	13. Find the number of arrangements of 3 letters of the word PRISM if letters may repeat.
4. RHOMBUS	14. Find the number of 3-digit numbers containing only 3s and 7s.
5. TWO	15. Find the number of 4-letter arrangements of all vowels including Y where a letter may repeat.
6. EQUATION	16. Find the number of 3-digit numbers with all digits odd.
7. TRAPEZOIDS	17. Find the number of 3-digit numbers with all digits even.
8. NUMBER	18. How many codes can be made with 2 letters followed by 3 digits if the letter "O" is not used?
9. TWICE	19. Find the number of 3-digit numbers without the digits 4 and 8.
10. SQUARE	20. Find the number of 4-digit even numbers without the digit 2.

Permutations 2

Find the number of arrangements of the letters of the word.

1. MISSISSIPPI	12. HELPLESS
2. QUEEN	13. SASSY
3. SPINNER	14. SHELLFISH
4. LOLLYPOP	15. POSSESS
5. BALL	16. GUERRILLA
6. DEEDED	17. HIGHER
7. GOOGOL	18. OVERWORK
8. TOMFOOL	19. PACIFIC
9. BEEKEEPER	20. PAYDAY
10. PUPPET	21. PARABOLA
11. RASPBERRY	22. NONSENSE

Permutations 3

Find the number of arrangements using multiplication principle, placing the restrictions first.

1. Find the number of ways to arrange 4 boys and 3 girls in a row with a boy at each end.	9. Find the number of ways to park 5 different colored cars with the red one in the middle.
2. Find the number of ways to arrange 4 boys and 3 girls in a row, alternating boys and girls.	10. Find the number of ways to park 5 different colored cars with the red one at either end.
3. Find the number of ways to arrange 4 boys and 3 girls in a row with the girls completely centered.	11. Find the number of ways to park 5 different colored cars with the red one next to the blue one.
4. Find the number of 5-letter codes that have a vowel (not Y) in the middle with no letter repeating.	12. Find the number of 5-digit numbers with the middle digit prime and no repeated digits.
5. Find the number of arrangements of the letters PRIME if the first letter must be a vowel.	13. Find the number of 5-digit odd numbers with the middle digit 7, first digit an odd prime, and no repeated digits.
6. Find the number of arrangements of the letters PRIME if the vowels must be first and last.	14. Find the number of ways to arrange the letters of SUPER with the first and last letters vowels.
7. Find the number of arrangements of the 10 digits if the first one must be odd and the last one must be even.	15. Find the number of 3-digit numbers that contain at least one 1.
8. Find the number of 3-digit even numbers selecting from 3, 4, 5, and 6 without repeating digits.	16. Find the number of batting orders of 9 baseball players with the pitcher last and the 3 basemen starting the lineup.

Permutations 4

Find the number of different:	Find the sum of:
1. choral groups, selecting 6 of 8 boys and 6 of 8 girls.	9. all 2-digit permutations of 12.
2. school groups, selecting 5 of 7 teachers and 4 of 6 principals.	10. all 3-digit permutations of 123.
3. bipartisan committees, selecting 3 of 5 Republicans and 3 of 5 Democrats.	11. all 4-digit permutations of 1234.
4. "words" formed by concatenating permutations of MAN to those of WOMAN.	12. all 3-digit permutations of 159.
5. "words" formed by concatenating permutations of MOM to those of GRAMM.	13. all 4-digit permutations of 1256.
6. debate teams, selecting 2 of 9 boys and 2 of 10 girls.	14. all 3-digit permutations of 234.
7. "words" formed by concatenating permutations of GAME to those of BOARDS.	15. all 3-digit permutations of 178.
8. "words" formed by concatenating permutations of SASS to those of GREIGG.	16. all 4-digit permutations of 1357.

Place Value 1

Write the numeral for the English expression.

1. six hundred ten thousand	12. six and one hundred-thousandth
2. six hundred and ten thousandths	13. sixty ten-thousandths
3. six ten-thousandths	14. sixty and six tenths
4. six hundred ten-thousandths	15. six hundred and six hundredths
5. six hundred ten thousandths	16. six thousand six ten-thousandths
6. six and ten thousandths	17. sixty and sixty hundredths
7. six hundred and one ten-thousandth	18. six hundred and sixty hundredths
8. six hundred thousand	19. sixty thousand six
9. six hundred thousandths	20. sixty and sixty-one thousandths
10. six hundred-thousandths	21. six and sixty-one ten-thousandths
11. six and one hundred thousandths	22. six thousand six hundred six

Place Value 2

Arrange the digits as specified, using each digit once.	Round to the specified decimal place.			
		10th	100th	1000th
1. digits 2, 3, 5, 6, 7, and 9 into two 3-digit numbers with least positive difference	12. 4.1146			
2. digits 1, 3, 6, 7, 8, and 9 into two 3-digit numbers with greatest positive difference	13. 3.5876			
3. digits 1, 2, 4, 6, 7, and 8 into two 3-digit numbers with least positive sum	14. 0.4392			
4. digits 2, 3, 5, 6, 8, and 9 into two 3-digit numbers with greatest positive sum	15. 1.1515			
5. digits 0, 1, 3, 4, 7, and 9 into two 3-digit numbers with least positive difference	16. 7.0629			
6. digits 0, 2, 4, 6, 8, and 9 into two 3-digit numbers with greatest positive difference	17. 6.2656			
7. digits 1, 3, 5, 7, 8, and 9 into two 3-digit numbers with least positive sum	18. 2.7717			
8. digits 1, 4, 5, 6, 7, and 8 into two 3-digit numbers with greatest positive sum	19. 4.3645			
9. digits 1, 2, 3, 5, 6, 7, 8, and 9 into the two 4-digit numbers with the least positive difference	20. 9.2288			
10. digits 0, 1, 3, 4, 5, 6, 7, and 9 into the two 4-digit numbers with the least positive difference	21. 4.5812			
11. digits 1, 2, 3, 4, 5, 6, 8, and 9 into the two 4-digit numbers with the greatest positive difference	22. 7.5555			

Polygons 1

Find the measure of one angle.	Find the number of sides.	Find the number of sides.
1. regular dodecagon	12. each angle = 168°	23. # of diagonals = 90
2. regular pentagon	13. each angle = 165°	24. # of diagonals = 20
3. regular 36-gon	14. each angle = 172°	25. # of diagonals = 54
4. regular decagon	15. each angle = 120°	26. # of diagonals = 35
5. regular pentadecagon	16. each angle = 177°	27. # of diagonals = 209
6. regular 18-gon	17. each angle = 135°	28. # of diagonals = 104
7. regular icosagon	18. each angle = 176°	29. # of diagonals = 170
8. regular 60-gon	19. each angle = 171°	30. # of diagonals = 275
9. regular 30-gon	20. each angle = 60°	31. # of diagonals = 495
10. regular octagon	21. each angle = 140°	32. # of diagonals = 27
11. regular 72-gon	22. each angle = 162°	33. # of diagonals = 44

Polygons 2

Find the sum of the angles in degrees.	Find the number of sides given the sum of the angles.
1. dodecagon	12. 2340°
2. 17-gon	13. 7200°
3. octagon	14. 6120°
4. hexagon	15. 3960°
5. decagon	16. 1260°
6. 14-gon	17. 1980°
7. 32-gon	18. 8460°
8. heptagon	19. 5040°
9. pentagon	20. 9000°
10. 11-gon	21. 4500°
11. 20-gon	22. 2520°

Polygons 3

Find the number of diagonals for the convex polygon.

1.	number of sides = 20	12.	each angle = 176°
2.	each angle = 108°	13.	sum of angles = 2520°
3.	sum of angles = 3600°	14.	each angle = 135°
4.	number of sides = 18	15.	each angle = 168°
5.	sum of angles = 900°	16.	number of sides = 16
6.	shape is dodecagon	17.	sum of angles = 7200°
7.	number of sides = 23	18.	each angle = 120°
8.	sum of angles = 1980°	19.	shape is pentadecagon
9.	number of sides = 40	20.	each angle = 165°
10.	shape is nonagon	21.	shape is decagon
11.	sum of angles = 2700°	22.	shape is icosagon

Polygons 4

Answer as indicated. Angles are in degrees.

1. The measures of the interior angles of a quadrilateral are x, 3x, 2x + 30, and 2x − 10. Find the sum of the least and the greatest angles.

2. The measures of the interior angles of a triangle are 5x, 4x + 8, and 3x − 2. Find the median of the angles.

3. The measures of the interior angles of a pentagon are x + 30, 4x − 10, 2x, 3x + 10, and 2x + 18. Find the angle measure that is a perfect square.

4. The measures of the interior angles of a quadrilateral are 4x, 3x − 1, 2x + 21, and 5x − 10. Find the median of the angles.

5. The measures of the interior angles of a hexagon are 3x, 4x − 7, 2x, 5x − 13, 6x − 15, and 7x − 55. Find the greatest angle.

6. One angle of a regular decagon is how much greater than one angle of a regular nonagon?

7. Find the measure of the smaller angle formed by a side and the nearest short diagonal of a regular pentagon.

8. Find the measure of the smaller angle formed by a side and the nearest short diagonal of a regular dodecagon.

9. The sum of the angles of a 22-gon is how much greater than the sum of the angles of a 14-gon?

10. The sum of the angles of a 54-gon is how much greater than the sum of the angles of a 53-gon?

11. Find the measure of the smaller angle formed by a side and the nearest short diagonal of a regular octagon.

12. One angle of a regular 16-gon is how much greater than one angle of a regular 15-gon?

Primes 1

Use divisibility rules to label the number prime, composite, or neither. If composite, state one factor other than 1 or the number. Answers may vary for the factors.

1. 1	17. 133	33. 411	49. 1001
2. 37	18. 134	34. 417	50. 1111
3. 51	19. 137	35. 451	51. 1221
4. 53	20. 143	36. 513	52. 1243
5. 55	21. 159	37. 517	53. 2361
6. 57	22. 187	38. 531	54. 2385
7. 59	23. 189	39. 561	55. 2403
8. 91	24. 201	40. 583	56. 3331
9. 93	25. 203	41. 671	57. 3773
10. 97	26. 207	42. 693	58. 4011
11. 101	27. 209	43. 707	59. 4527
12. 111	28. 253	44. 781	60. 6111
13. 113	29. 313	45. 801	61. 7707
14. 119	30. 321	46. 811	62. 8541
15. 121	31. 327	47. 931	63. 9001
16. 123	32. 329	48. 957	64. 9647

Primes 2

Write the prime factorization in rows (no trees), each equivalent to the original number.
The final row must be all primes in ascending order with exponents.

1. 108	9. 500	17. 1260
2. 144	10. 576	18. 1625
3. 189	11. 606	19. 1980
4. 216	12. 621	20. 2025
5. 297	13. 675	21. 2106
6. 333	14. 720	22. 5400
7. 360	15. 840	23. 6000
8. 375	16. 950	24. 6600

Primes 3

Answer as indicated.

1. Find the sum of the four numbers: the least 2-digit twin primes and the greatest 2-digit twin primes.	9. How many 2-digit primes may be formed by selecting two digits from {2, 3, 4, 7}?
2. Find the sum of the primes from 80 to 100.	10. Find the sum of the primes from 100 to 120.
3. How many pairs of twin primes are greater than 100 and less than 200?	11. Find the three least 3-digit numbers with exactly 3 factors each.
4. Find the number of primes between 40 and 80.	12. Find the least 3-digit prime with a prime digit sum.
5. Find the least positive integer divisible by 4 primes.	13. Find the greatest 3-digit prime with a prime digit sum.
6. Find all 2-digit pairs of numbers such that both the number and the number with the digits reversed are primes.	14. Find the product of the primes between 20 and 30.
7. Find the greatest 3-digit positive integer divisible by the first 3 primes.	15. Use the digits 2, 3, 7, and 9 each once to create two primes with the greatest product.
8. Find the sum of the prime factors of 21,945.	16. Find the sum of the least 2-digit prime and the greatest 2-digit prime, both with a prime digit sum.

Primes 4

Answer as indicated.

1.	Find the sum of the positive differences of all pairs of twin primes less than 100.	8.	If p and p+3 are both primes, find the least composite number not divisible by either one.
2.	Find the sum x + y where x = the sum of the first 5 primes and y = the sum of the reciprocals of the first 3 primes.	9.	Find the sum of the least 2-digit prime and the greatest 2-digit prime, both with the product of its digits prime.
3.	Selecting two of the first eight primes, find the probability of their sum equaling 16.	10.	Find five primes that form an arithmetic sequence with a common difference of 6.
4.	How many ordered triples of different primes have the sum of their elements equal to 36?	11.	When the prime 97 is written as the sum of one composite and one prime, find the least possible positive difference of the two numbers.
5.	How many of the first 20 positive integers may be written as the sum of 2 distinct primes?	12.	Find the sum of the greatest 3-digit number with exactly 3 factors and the least 4-digit number with exactly 3 factors.
6.	How many ordered triples of primes have the sum of their elements equal to 33?	13.	How many of the first 20 positive integers may be written as the sum of 3 distinct primes?
7.	The sum of the least three 3-digit primes is how much greater than the sum of the greatest three 2-digit primes?	14.	Selecting two of the first twelve primes, find the probability of their sum equaling 36.

Prisms 1

Find the volume in cubic units of the rectangular prism with the given unit dimensions. Use mental math.

1. 4, 5, 8	20. 8, 8, 20
2. 2, 7, 10	21. 7, 8, 11
3. 5, 6, 10	22. 30, 35, 40
4. 10, 12, 20	23. 8, 25, 30
5. 3, 9, 11	24. 4, 15, 15
6. 8, 9, 10	25. 5, 19, 20
7. 10, 20, 30	26. 20, 35, 60
8. 3, 4, 6	27. 7, 10, 12
9. 5, 5, 5	28. 1, 11, 17
10. 4, 6, 11	29. 50, 60, 90
11. 10, 12, 12	30. 2, 7, 25
12. 5, 7, 11	31. 20, 20, 24
13. 40, 50, 60	32. 2.1, 8, 10
14. 9, 9, 9	33. 1.1, 1.1, 200
15. 8, 15, 30	34. 2.5, 4, 13
16. 2, 4, 7	35. 2, 6.1, 20
17. 2, 9, 15	36. 1.5, 4, 15
18. 2, 15, 15	37. 2.5, 3, 10
19. 4, 6, 6	38. 1.3, 1.3, 100

Prisms 2

Find the surface area of the rectangular prism in square units. Show pairwise work.

1. 11, 15, 30	9. 10, 20, 30
2. 4.5, 11, 13	10. 4, 11, 11
3. 5, 12, 30	11. 30, 40, 60
4. 3.5, 6, 7	12. 2.5, 8, 10
5. 2.5, 3, 10	13. 1.5, 4, 15
6. 6, 10, 12	14. 1.2, 1.5, 10
7. 10, 15, 50	15. 5, 6.3, 10
8. 10, 12, 20	16. 2.5, 8, 9

Prisms 3

Find the volume in cubic units of the rectangular prism given areas of three faces in square units.

1. 24, 32, 48	7. 72, 120, 135	13. 32, 40, 80
2. 30, 40, 48	8. 35, 45, 63	14. 90, 99, 110
3. 30, 35, 42	9. 88, 104, 143	15. 63, 70, 90
4. 24, 48, 72	10. 17.5, 35, 24.5	16. 54, 72, 108
5. 6, 22, 33	11. 15, 51, 85	17. 22.5, 31.5, 35
6. 24, 48, 128	12. 48, 54, 72	18. 88, 96, 132

Prisms 4

Find the space diagonal of the rectangular prism with the given unit dimensions.

1. 3, 4, 5	12. 2, 3, $2\sqrt{3}$
2. 6, 8, 10	13. 1, 5, $2\sqrt{7}$
3. 1, 5, 10	14. 4, 9, $2\sqrt{7}$
4. 5, 10, 15	15. $\sqrt{5}, \sqrt{7}, \sqrt{13}$
5. 8, 9, 12	16. 3, 4, $2\sqrt{6}$
6. 3, 5, 8	17. 2, 6, $4\sqrt{2}$
7. 1, 2, 2	18. 1, 3, $5\sqrt{5}$
8. 2, 4, 13	19. 5, 8, $3\sqrt{7}$
9. 7, 24, 60	20. 1, 3, $5\sqrt{6}$
10. 2, 6, 9	21. 1, 5, $7\sqrt{2}$
11. 4, 5, 11	22. $\sqrt{3}, \sqrt{7}, \sqrt{71}$

Prisms 5

Find the volume in cubic units of the rectangular prism given areas of three faces in square units.

1. 22, 24, 33	7. 30, 40, 48	13. 24, 48, 72
2. 24, 48, 200	8. 17, 51, 108	14. 19.5, 39, 128
3. 17.5, 35, 50	9. 20, 70, 126	15. 18, 34, 68
4. 20, 50, 160	10. 20, 35, 112	16. 10, 15, 54
5. 13, 26, 98	11. 12, 75, 121	17. 20, 24, 120
6. 52, 65, 500	12. 30, 60, 72	18. 24, 30, 45

Prisms 6

Find the volume of the prism in cubic units with height L of the base, area A of the base, and altitude h of the prism.

1.	triangular prism sides of base 16, 30, 34; h = 40	8.	trapezoidal prism b = 5, B = 20, L = 18, h = 22
2.	regular hexagonal prism side of base 10, h = 15	9.	pentagonal prism A = 42, h = 9
3.	rhomboidal prism diagonals of base 10 and 15, h = 8	10.	isosceles triangular prism sides of base 50, 50, 28; h = 100
4.	octagonal prism A = 75, h = 11	11.	equilateral triangular prism side of base 18, h = 10
5.	equilateral triangular prism side of base 22, h = 5	12.	regular hexagonal prism side of base 14, h = 20
6.	isosceles triangular prism sides of base 26, 26, 20; h = 30	13.	rhomboidal prism diagonals of base 25 and 40, h = 15
7.	trapezoidal prism b = 6, B = 14, L = 5, h = 11	14.	triangular prism sides of base 11, 60, 61; h = 20

Prisms 7

A slice is made along the internal diagonal plane of a rectangular prism with the given dimensions to create a wedge. Find the volume and surface area of the wedge. NTS

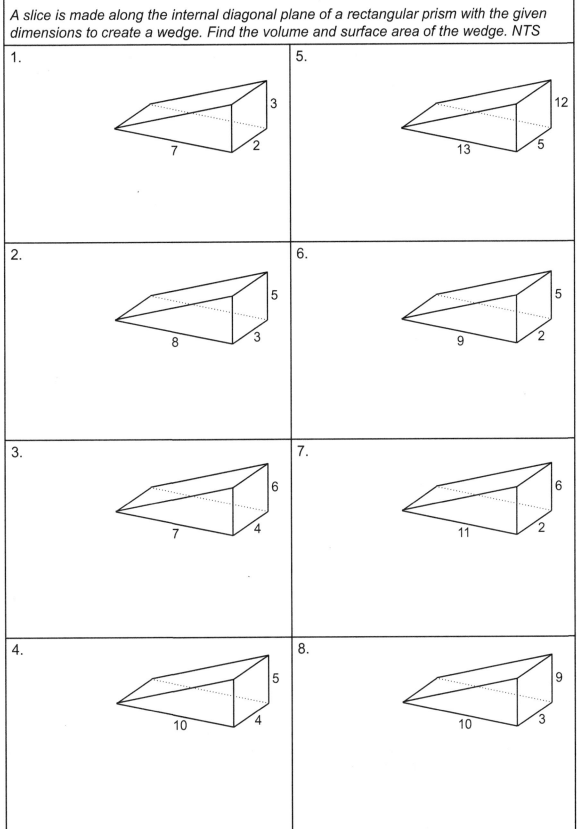

1.

3
7   2

5.

12
13   5

2.

5
8   3

6.

5
9   2

3.

6
7   4

7.

6
11   2

4.

5
10   4

8.

9
10   3

Prisms 8

Answer as indicated for the rectangular prism. Measures are units, square units, and cubic units unless otherwise specified.

1. The dimensions of a rectangular prism are 3 distinct whole numbers greater than 1. The volume is 231. Find the surface area.

2. A rectangular box open at the top has base 10 by 20 and height 30. Find the inside plus outside surface area.

3. The dimensions of a rectangular prism are 3 distinct whole numbers greater than 1. The volume is 715. Find the surface area.

4. A rectangular box open at the top has base 5 by 6 and height 11. Find the inside plus outside surface area.

5. The dimensions of a rectangular prism are 1, 2, and x. The space diagonal is 3. Find x.

6. Of 2 rectangular prisms, one has volume 30 and the second 105. Find the ratio of their non-congruent sides if 2 sides are congruent and all are greater than 1.

7. A rectangular box measures 2 by 4 by 5. Find the length of the longest stick that can fit in the box.

8. The dimensions of a rectangular prism are 1, 6, and x. The space diagonal is 19. Find x.

9. The dimensions of a rectangular prism are 2, 5, and x. The space diagonal is 15. Find x.

10. Find the maximum number of rectangular blocks 3 by 5 by 2 that can fit into a box 15 by 30 by 12.

11. Find the maximum number of rectangular blocks 6" by 4" by 8" that can fit into a box 1.5' by 1' by 2'.

12. Find the diagonal of a rectangular prism with volume 24 and height and width both $2\sqrt{2}$.

Prisms 9

Given the base of a prism, find the sum of the number of edges, vertices, and faces.	*Find the missing value for the prism using the formulae relating E, F, and V.*
1. triangle	14. 15 edges _____ vertices
2. rectangle	15. 18 edges _____ faces
3. pentagon	16. 9 faces _____ vertices
4. hexagon	17. 13 faces _____ edges
5. heptagon	18. 8 vertices _____ edges
6. octagon	19. 16 vertices _____ faces
7. nonagon	20. 27 edges _____ vertices
8. decagon	21. 18 faces _____ edges
9. 11-gon	22. 24 vertices _____ faces
10. dodecagon	23. 30 edges _____ faces
11. pentadecagon	24. 15 faces _____ vertices
12. 17-gon	25. 30 vertices _____ edges
13. 21-gon	26. 9 edges _____ faces

Prisms 10

Answer as indicated. All tanks and trays are rectangular prisms. Trays are formed by cutting squares from corners of a rectangle and folding up the sides.

1. A tall tank partially full of water has a base 14 inches by 2.5 feet. When a rock totally sinks in the water, the water rises .3 inches. Find the volume of the rock in cubic inches.

2. A tall tank partially full of water has a base 3 feet by 3 feet. When a heavy, solid cube 1 foot on an edge totally sinks in the water, how many inches will the water level rise?

3. A tall tank partially full of water has a base 2.25 feet by 1.25 feet. When a rock totally sinks in the water, the water rises .6 inches. Find the volume of the rock in cubic inches.

4. A tall tank partially full of water has a base 4 feet by 9 feet. A cube with edge 1.5 feet sinks in the water. How many inches will the water level rise?

5. A tank with base 3 by 6 feet is 4 feet tall and 7/8 full of water. A cube with edge 2 feet sinks in the water. Find the number of inches from the top of the water level to the top of the tank.

6. Two-inch squares were cut from the corners of a square with area 121 square inches. Find the volume of the tray.

7. One-inch squares were cut from the corners of a rectangle. The tray has volume 75 cubic inches and length triple the width. Find the dimensions of the rectangle.

8. Three-inch squares were cut from the corners of a square. Find the interior surface area in square inches of the tray with volume 432 cubic inches.

9. Four-inch squares were cut from the corners of a square. The tray has volume 256 cubic inches. Find its total surface area, inside and outside.

10. Five-inch squares were cut from the corners of a rectangle. The tray has volume 880 cubic inches and length 5 more than the width. Find the dimensions of the rectangle.

Probability 1

Rolling a standard die, find the probability of:	Rolling a standard die, find the odds of:	Drawing once from a card deck, find the odds of:
1. a 6	17. a 6	33. ace of spades
2. a prime number	18. a prime number	34. ace
3. a 7	19. a 7	35. club
4. an even number	20. an even number	36. red card
5. a perfect square	21. a perfect square	37. a perfect square
6. a number less than 6	22. a number less than 6	38. a number less than 6
7. a number greater than 2	23. a number greater than 2	39. even card
8. a 2-digit number	24. a 2-digit number	40. odd card
9. a factor of 6	25. a factor of 6	41. face card
10. a multiple of 5	26. a multiple of 5	42. multiple of 5
11. a number less than 7	27. a number less than 7	43. multiple of 4
12. an odd number	28. an odd number	44. multiple of 3
13. a multiple of 3	29. a multiple of 3	45. multiple of 2
14. a perfect cube	30. a perfect cube	46. a perfect cube
15. a composite number	31. a composite number	47. heart
16. a factor of 10	32. a factor of 10	48. queen of hearts

302

Probability 2

Find the probability of choosing a point at random from the shaded area given that the point lies inside the outer border. One box equals one square unit.

1.

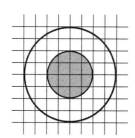

2.

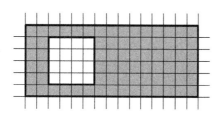

3.

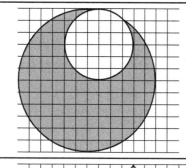

4.

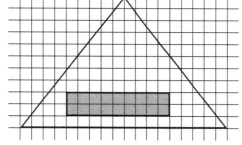

5.

6.

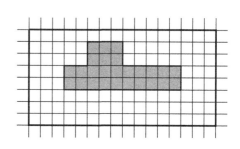

7.

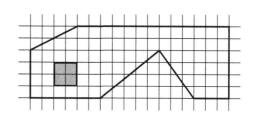

8.

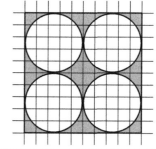

9.

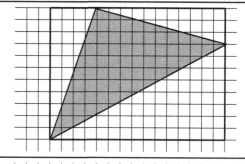

10.

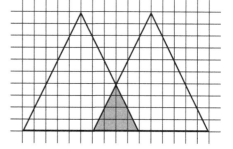

Probability 3

Form each probability by "make-a-list" over "multiplication principle." Assume solids are numbered consecutively beginning with 1.

1.	Find the probability that the face-up sum is prime when rolling two dice.	6.	Rolling two regular 20-gons (1–20), find the probability that the face-up sum equals 37.
2.	Rolling three octahedra (1–8), find the probability that the face-up sum equals 5.	7.	Rolling two tetrahedra (1–4), find the probability that the face-down sum is 5.
3.	Rolling 2 dodecahedra (1–12), find the probability that the face-up sum is 22.	8.	When flipping 4 coins, find the probability of exactly 2 heads and 2 tails.
4.	Find the probability that the face-up sum is a perfect cube when rolling two standard dice.	9.	When flipping 5 coins, find the probability of exactly 2 heads.
5.	Find the probability that the face-up sum is 16 when rolling three standard dice.	10.	Find the probability that the face-up sum is 7 when rolling two standard dice.

Probability 4

Find the probability when rolling two dice, following the meaning of "OR."	*A bowl contains 13 blue, 11 yellow, 14 red, and 12 green marbles. Find the probability.*
1. The face-up sum equals 7 or 11.	9. P(blue OR yellow)
2. The face-up sum equals 4 or 6.	10. P(yellow OR red)
3. The face-up sum equals an even prime or an odd composite.	11. P(red OR green)
4. The face-up sum equals a perfect square.	12. P(green OR blue)
5. The face-up sum is greater than 9.	13. P(blue OR yellow OR red)
6. The face-up sum equals an odd prime.	14. P(yellow OR green OR blue)
7. The face-up sum equals 3 or an even 2-digit number.	15. P(blue OR green OR red)
8. The face-up sum equals a perfect cube or an odd 2-digit number.	16. P(red OR yellow OR green)

Probability 5

Find the probability following OR and subtracting the overlap, drawing once:

from a standard deck of cards.	*from 26 letters of the alphabet (Y is vowel).*
1. P(ace OR red)	12. P(vowel OR even ordinal letter)
2. P(queen OR heart)	13. P(vowel OR odd ordinal letter)
3. P(face OR spade)	14. P(consonant OR even ordinal letter)
4. P (one-digit prime OR red)	15. P(consonant OR odd ordinal letter)
5. P(8 OR king OR black)	16. P(vowel OR in word HOUSE)
6. P(one-digit even OR diamond)	17. P(consonant OR in word HOUSE)
7. P(perfect square OR club)	18. P(in word HOME OR in word HOUSE)
8. P(diamond OR red)	19. P(vowel OR in word MOVE)
9. P(black OR club)	20. P(consonant OR in word MOVE)
10. P(king OR face card)	21. P(in word MOVE OR in word CARE)
11. P (two-digit prime OR black)	22. P(in word TIME OR in word GREAT)

Probability 6

Answer as indicated, following the meaning of "AND" in probability.

1. The probability of rain is 3/4 and of forgetting one's umbrella is 2/3. Find P(rain AND have umbrella).

2. Find the probability of having 3 girls as children assuming P(girl) = P(boy).

3. Find the probability of Team A losing 2 consecutive games if it has a .75 chance of winning any game.

4. Find the probability of guessing correctly on four consecutive true-false questions.

5. Draw one card from a standard deck and roll one die. Find P(heart AND prime).

6. Find the probability of getting a head first and a tail second when tossing a coin twice.

7. On 2 draws with replacement from the natural numbers less than 21, find P(odd, multiple of 4).

8. Find the probability of completing a maze on the first try if the maze has 8 forks and each fork goes right or left.

9. Draw one card from a standard deck and roll one die. Find P(king AND odd).

10. Find the probability of getting HTTHT when tossing a coin five times.

11. Draw one card from a standard deck and roll one die. Find P(face card AND composite).

12. Bob's foul shot success is 70%. Find the probability of his missing his next 3 foul shots.

13. Draw one card from a standard deck and roll one die. Find P(black AND perfect square).

14. Find the probability of sunken souffles on Geena's next four baked if the success of her souffle rising is 80%.

Probability 7

A bowl contains 10 green, 14 yellow, 16 red, and 25 blue marbles. Find the specified probability for successive draws without replacement.

1. P(R, R)	12. P(B, R, G)
2. P(B, R)	13. P(NOT Y, NOT Y, NOT Y)
3. P(Y, Y)	14. P(B OR G, R OR Y)
4. P(NOT B, NOT B)	15. P(B OR Y, R OR G)
5. P(R, B, Y)	16. P(NOT G, NOT G)
6. P(G, G, R)	17. P(B, NOT B, NOT B)
7. P(B, B, Y)	18. P(R, R, R)
8. P(R, Y, R)	19. P(Y, Y, Y)
9. P(R, Y, G)	20. P(G, G, G)
10. P(NOT B, NOT B, NOT B)	21. P(R, B, R)
11. P(Y, Y, R)	22. P(G, G, G, G)

Probability 8

For the "at least" situation, find the probability of:

1.	at least one of two people born on Monday.	8.	at least one tail when flipping a coin 3 times.
2.	at least one "4" when rolling a die 3 times.	9.	at least 1 club when drawing 2 cards from a deck without replacement.
3.	making at least one goal in 3 tries if a player makes 1/4 of his goals.	10.	making at least one goal in 3 tries if a player makes 1/3 of his goals.
4.	at least 1 head when flipping a coin 4 times.	11.	rain at least 1 day in the next 2 days if rain occurs 2 of 7 days.
5.	at least one of 3 people born on Friday.	12.	at least 1 black card when drawing 2 cards from a standard deck without replacement.
6.	sun at least 1 day in the next 3 days if the sun shines 3 of 7 days.	13.	at least 1 hit in 3 tries if the probability of a hit is 3/4.
7.	at least 1 blue, selecting 2 random balls without replacement from 4 blue, 3 green, and 3 pink.	14.	at least one red, selecting 2 random balls without replacement from 4 blue, 2 white, and 3 red.

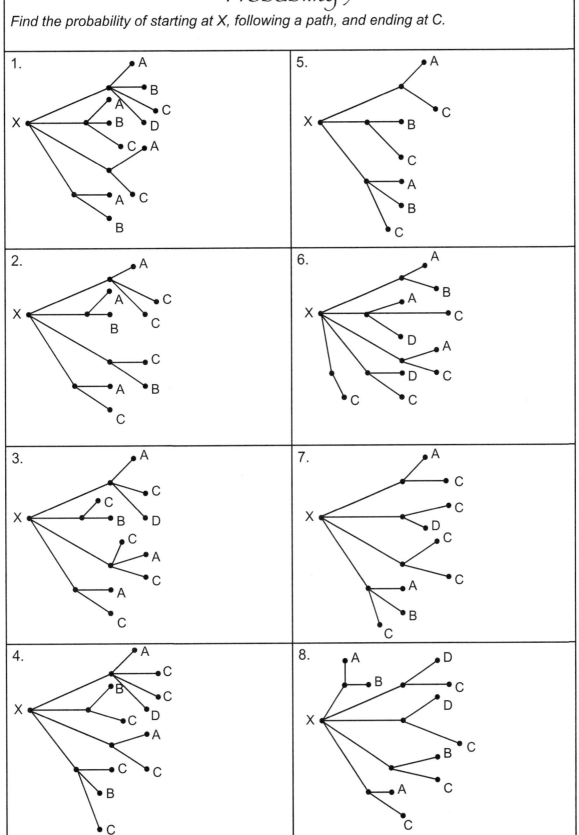

Probability 9

Find the probability of starting at X, following a path, and ending at C.

Probability 10

Find the probability in a variety of settings.

Find the probability using a combination as the denominator.

1. Select one of the divisors of 720. Find the probability that it is divisible by 4.

7. A bag contains 1Q, 1D, 1N, and 1P. Pull 2 coins without replacement. Find P(greater than 11¢).

2. Find probability that among all permutations of GREED, the letters are in alphabetical order.

8. Select 2 distinct numbers from the set {3, 4, 5, 6, 7}. Find the probability that their product is even.

3. Find the probability that a natural number ≤ 40 is relatively prime to 40.

9. Select 3 distinct numbers from the set {1, 2, 3, 4, 5, 6}. Find the probability that their sum is odd.

4. Select one fraction from among 1/2, 2/3, 3/4, 4/5, 5/6, 6/7, 7/8, and 8/9. Find the probability that it terminates.

10. A jar contains 2D, 2N, and 2P. Pull 3 coins without replacement. Find P(less than 17¢).

5. Let a and b be whole numbers such that 1≤a≤4 and 1≤b≤10. Find the probability that a and b are both odd.

11. Select 2 distinct digits. Find the probability that their positive difference is 4.

6. Randomly select a multiple of 9 between 0 and 75. Randomly select a multiple of 11 between 35 and 100. Find the probability that the two numbers are the same.

12. Select 2 of the prime numbers less than 35. Find the probability that they are twin.

Probability 11

Find the overall probability by adding individual probabilities.	*Find the probability with a "given" in one event (no prior dependence).*
1. Rolling a die twice, find the probability that the second roll is greater than the first.	7. Given that when tossing 3 coins, at least 1 is a head. Find P(at least 1 tail).
2. Jar A has 8 red and 2 blue marbles. Jar B has 5 blue and 7 pink marbles. Draw once from Jar A. If red, draw a 2nd from A; but if blue, draw a 2nd from B. Find P(1st any, 2nd not pink).	8. Given that when tossing 3 coins, exactly 1 is a head. Find P(2 successive tails).
3. Rolling a die twice, find the probability that the second roll is less than or equal to the first.	9. A jar contains 6 red, 4 blue, 7 green, and 3 pink marbles. Draw 1 randomly. Given that it is not blue, find the probability of red.
4. Jar A has 8 tan and 8 red marbles. Jar B has 10 red and 20 green marbles. Draw once from Jar A. If red, draw a 2nd from A; but if tan, draw a 2nd from B. Find P(1st any, 2nd not green).	10. Roll a red and blue die together. Given that the sum of the numbers is 10, find P(red 4).
5. A packet contains 7 seeds: 2 produce blue flowers, 3 white, and 2 red. Select 2 seeds randomly. Find P(same color).	11. Roll a red and blue die together. Let P1 = P(red die even and blue die 4). Let P2 = P(red die even given blue die 4). Find P1 + P2.
6. A packet contains 9 seeds: 3 produce blue flowers, 4 white, and 2 red. Select 2 seeds randomly. Find P(same color).	12. Roll a red and blue die together. Given that the sum of the numbers is 7, find P(red 4).

Probability 12

Find the probability that 3 numbers selected randomly form the sides of an acute Δ.	*Find the probability that 3 numbers selected randomly form the angles of an acute Δ.*
1. 5, 8, 12, 13	9. 20, 30, 40, 50, 60, 70, and 80
2. 3, 4, 5, 8	10. 35, 50, 55, 60, 65, and 80
3. 5, 9, 11, 12	11. 35, 50, 60, 70, and 85
4. 6, 8, 9, 10	12. 20, 25, 35, 65, 70, 75, 80, and 85
5. 5, 5, 6, 7	13. 14, 34, 54, 56, 60, 70, 76, 80, and 86
6. 7, 8, 15, 17	14. 27, 43, 50, 57, 67, 73, and 80
7. 12, 14, 16, 20	15. 15, 25, 35, 45, 55, 60, 65, 80, and 85
8. 6, 7, 8, 11	16. 31, 49, 50, 51, 61, 70, 79, and 80

Problem Solving 1

Solve by working backwards.	*Solve by making a chart.*
1. Five boys with initials A, B, C, D, and E exchanged cards. Only A had cards to start. A gave B 1/3 of his cards plus 3. B gave C 1/3 of his cards plus 3. C gave D 1/3 of his cards plus 3. D gave E 15 cards. How many cards did A have to start?	5. A 6-cup mixture is 1/4 wheat flour and 3/4 white flour. Add 4 cups of wheat flour. The new 10-cup mixture is what percent wheat flour?
2. Find the least natural number such that the operation cycle can occur 3 times, always yielding natural numbers: subtract 1, take 2/3 of the difference.	6. Four women--Amy, Belle, Cleo, and Dee--visited Edie one day (either AM or PM). Amy visited at 8:00, Belle at 9:00, Cleo at 10:00, and Dee at 11:00. Cleo did not visit between Belle and Dee. At least one visited between Amy and Belle. Amy did not visit before Dee. Who visited last?
3. Certain micro-organisms contained in a jar quadruple every 30 seconds. After 1.5 minutes, the jar contained 2112 of them. How many were in the jar to start?	7. A uniform rate of 720 ticks per cycle and 20 cycles per hour is the same rate as what number of ticks per second?
4. Start with a number. Add 5, minus 8, add 3, minus 6, add 2, minus 4, and add 7. The result is 14. What was the starting number?	8. Saving money at a uniform weekly rate yielded $78 savings in 6 weeks. At the same rate, in how many more weeks will the savings be $208?

Problem Solving 2

Solve each problem using the Trial and Error (combined with reasoning) method.	*Solve by introducing a line.*
1. A bag contains 11 red, 6 blue, and 5 green marbles. Find the least number of blue marbles that must be added so that the probability of drawing one blue marble is greater than 1/2.	6. The area of $\triangle ABC$ is 36 square units. Find the area of $\triangle BCD$. (NTS)
2. A game has 5 players, each scoring from 0 to 100 inclusive. Their mean score is 80. Find the greatest number of players who can score exactly 50.	7. Find the measure of an interior angle of a regular 5-pointed star.
3. From the digits 3, 4, 5, and 6, form the 3-digit number and the 1-digit number that yield the greatest product. Use each digit only once.	8. A square with area 16 is inscribed in one quarter of a circle. Find the area of the circle.
4. The sum of the digits of a certain two-digit number equals the square root of the number. Find the number.	9. Find the perimeter of the pentagon.
5. Place each of 2, 3, 4, 6, and 8 in one of the circles so that the vertical and horizontal products are equal.	10. Find the area of the quadrilateral, which is the intersection of 2 squares. C is the center of the smaller square.

Problem Solving 3

Solve by drawing a picture.

1. Cut a rectanglular paper in two. Cut one of the pieces in two. Find the sum of the *different* numbers of edges from all ways of creating the 3 detached polygons.

5. The area of the triangle formed by connecting midpoints of adjacent sides of a square is what fractional part of the area of the square?

2. 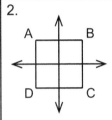 Flip the square centered at the origin 180° about the x-axis. Then flip it 180° about the y-axis. Find the final locations of the vertices.

6.  Find the greatest number of non-overlapping shaded regions when dividing the figure with exactly 2 straight lines.

3. There are 4 pieces of chain, each 3 links long. What is the least number of opens and closes of links to join all 12 links into one complete circle?

7. A, B, C, and D are 4 distnct points on a circle such that AB = BC = CD = AD. Name 2 diameters.

4. A, B, and C are jars with volumes in the ratio 2:1:3. Jar A is 3/4 full. Jars B and C are empty. Pour from A to fill B. Pour the remainder from A into C. Jar C is what fractional part full?

8. Fifteen points numbered consecutively 1 to 15 are on a circle. Cross off #3, #6, #9, and so on--every 3rd point not already crossed off. Find the number of the last point crossed off.

Problem Solving 4

Solve by making a list.

1. Find the number of right triangles in the square image. Assume trisection points.

2. How many distinct values are possible for 6 + 3 x 8 ÷ 2 by placing 1 pair of parentheses in the expression?

3. Randomly choose 3 of the numbers 3, 4, 5, 6, 8, 10. Find the probability that the numbers form sides of a triangle.

4. How many whole numbers between 23,000 and 23,600 have their digits in strictly ascending order?

5. N and D are positive digits. Find the number of simplified, unique, fractions N/D that are less than 1.

6. Switching only pairs of adjacent digits, find the least number of switches to change 12345 to 54321.

7. Connect the points (3, 7) and (11, 23) with a segment. How many lattice points are on the segment?

8. Randomly choose 4 of the numbers 3, 4, 5, 10, 18. Find the probability that the numbers form sides of a quadrilateral.

9. How many 3-digit numbers of the form AB1 are divisible by 3?

10. How many 3-digit numbers from 100 to 150 inclusive are divisible by each of their nonzero digits?

Problem Solving 5

Solve by making up numbers without loss of generality (WLOG).

1. If the natural number n is divided by 9, the remainder is 5. Find the remainder when 3n is divided by 9.	6. A flight occurred at 300 mph with the return trip on the exact route at 600 mph. Find the average speed for the round trip.
2. Double the length of one side of a square and halve the other. Find the ratio of the perimeters: original square to resulting rectangle.	7. Find the percent decrease in the area of a square if each side is decreased by 10%.
3. By what % is the area of a circle increased if its radius is increased by 20%?	8. The variable w increased by 10% of w is x. Then, x decreased by 50% of x is y. Finally, y increased by 20% of y is z. Find the percent that z is of w.
4. The set X contains only 3 consecutive natural numbers. The set Y contains only the next 3 consecutive natural numbers. Find the number of different values formed by summing an element of X and an element of Y.	9. Two sets of five consecutive natural numbers have exactly two numbers in common. Find the positive difference of the sums of the two sets.
5. AB = BC = CD = DE = DF. If a circle has a radius equal to (1/4)(AF) and point A is on the circle, then another point on the circle is between which two named points? A B C D E F	10. Find the percent change in the area of a square if one side is increased by 30% and one side is decreased by 30%.

Problem Solving 6

Solve by finding a pattern.

---

1. Find the number of pairs of vertical angles formed when 15 lines intersect in one point.

6. Following the pattern, find the sum of the terms in Row 15.
Row 1: 2 + 4
Row 2: 2 + 4 + 6
Row 3: 2 + 4 + 6 + 8

---

2. Evaluate:

$$1 \frac{1}{2} \times 1 \frac{1}{3} \times 1 \frac{1}{4} \times \ldots \times 1 \frac{1}{41}$$

7. Find the number of circles in the 15th row if the pattern continues.

---

3. How many dots are in the 10th image in the pattern?

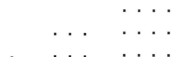

8. Find the number of total rectangles (including squares) in the 15th figure if the pattern continues.

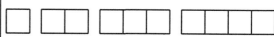

---

4. Find the rightmost digit of the 10th row.

```
        1
      2  3  4
    5  6  7  8  9
 10 11 12 13 14 15 16
```

9. Find the number of individual small squares in the 15th figure if the pattern continues.

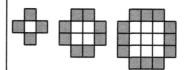

---

5. To make TUKs, 1 TUK uses 2 TEMs, 2 TUKs use 6 TEMs, 3 TUKs use 12 TEMs, and 4 TUKs use 20 TEMs. One hundred TUKs use how many TEMs?

10. Evaluate.

$$\frac{1}{2^1} + \frac{1}{2^2} + \frac{1}{2^3} + \ldots + \frac{1}{2^{14}} + \frac{1}{2^{15}}$$

Problem Solving 7

Solve by making a simpler problem.	*Solve by using logical reasoning.*
1. Find the sum of the first 12 positive perfect cubes.	6. Arrange the letters A, B, C, D, E, and F according to the following rules: 1. The word CAB appears. 2. D and E are at the ends in one of the two orders. 3. Two letters are beween A and E. 4. C is not next to D.
2. How many dots form the perimeter of an array of dots 799 dots by 800 dots?	7. Evaluate. Answer in exponential form. $(994-991)(991-988)(988-985)\ldots(7-4)(4-1)$
3. How many triangles are in the figure? 	8. Move and/or turn 2 of the short line segments shown to result in 4 of the small-size squares.
4. An 11 by 11 checkerboard is made up of alternating black and white squares with a black square in the bottom right corner. What fraction of the squares are black?	9. Find the next number in the sequence: 1, 1, 2, 3, 6, 9, 15, 135, 150, 285, ___
5. Find the value of $10^{2x} - 10^x + 1$ when x = 10.	10. If certain people only tell the truth on Sunday, Tuesday, and Thursday, then one of those people can say "I told the truth yesterday" on which day of the week?

Problem Solving 8

Write and solve one or more equations.

1. The denominator of a fraction is 1 less than twice the numerator. Subtracting 1 from the denominator, the fraction is 2/3. Find the original fraction.	5. The denominator of a fraction is 7 more than the numerator. Adding 5 to both, the fraction becomes 1/2. Find the original fraction.
2. The units digit of a number is 4 times the tens digit. Adding 54, the digits swap places. Find the number.	6. How many ounces of pure salt must be added to 5 ounces of a 20% salt solution to yield a 25% salt solution?
3. The numerator of a fraction is 3 less than the denominator. Subtracting 1 from both, the fraction is 1/2. Find the original fraction.	7. How many pounds of water must evaporate from 9 pounds of a 50% salt solution to yield a 75% salt solution?
4. A number is 6 times the sum of its digits. The units digit is 1 less than the tens digit. Find the number.	8. The sum of the digits of a number is 6. Subtracting 36 from the number, the digits swap places. Find the number.

Properties 1

Name the property illustrated. Choose among: ClPA, APA, IdPA, InPA, CPA, ClPM, APM, IdPM, InPM, CPM, DPMA, RPE, SPE, TPE, ZPM, APE, MPE.

1. 5 + 3 = 3 + 5	20. 5.2 x 1.7 = 1.7 x 5.2
2. If 1 + 1 = 2, then 1 + 1 + 5 = 2 + 5	21. If 6 + 5 = 11, then 11 = 6 + 5
3. (2 x 3) x 8 = 2 x (3 x 8)	22. (7 + 2) 5 = 35 + 10
4. 6 x (1/6) = 1	23. 0 + 13.2 = 13.2
5. 126 = 126	24. 2.1 x 3.3 is a unique real number
6. 3 + 4 is a unique real number	25. If 4 = 9 − 5 and 9 − 5 = 7 − 3, then 4 = 7 − 3
7. 17 + 0 = 17	26. 4 + (6 + 7) = (4 + 6) + 7
8. 4(5 + 3) = 20 + 12	27. 1 x 5.43 = 5.43
9. If 2 + 1 = 3, then 3 = 2 + 1	28. (7 x 2) x 5 = 7 x (2 x 5)
10. 2 x 3 is a unique real number	29. 9 + 5 = 5 + 9
11. 5 x 19 = 19 x 5	30. 67.4 = 67.4
12. 5 + (−5) = 0	31. 45.8 x 0 = 0
13. 7 + (6 + 1) = (7 + 6) + 1	32. If 2 + 3 = 5, then 4(2 + 3) = 4 x 5
14. 32 x 1 = 32	33. 0 = (−3.4) + 3.4
15. If 1 = 9 − 8 and 9 − 8 = 7 − 6, then 1 = 7 − 6	34. 3.1 + 4.56 is a unique real number
16. 0 x 999 = 0	35. (2/3) x (3/2) = 1
17. If 12 = 3x4, then 12x5 = 3x4x5.	36. If 12 = 11+1, then 11+1 = 12.
18. 6(2 + 4) = 6x2 + 6x4	37. 1 x 101 = 101
19. If 5x3 = 15, then 15 = 5x3.	38. 11 x (7 x 4) = (11 x 7) x 4

Properties 2

Name the property illustrated. Choose among: ClPA, APA, IdPA, InPA, CPA, ClPM,
APM, IdPM, InPM, CPM, DPMA, RPE, SPE, TPE, ZPM, APE, MPE.

1. $ab + 8 = ba + 8$	17. $0 + mn = mn$
2. If $xy = a$, then $xy + 4 = a + 4$	18. If $w = bc$ and $bc = x$, then $w = x$
3. $a(b + c) = ab + ac$	19. $3a \cdot 7b$ is a unique real number
4. $(abc)(1) = abc$	20. $4(ab) = 4(ba)$
5. $3x + 0 = 3x$	21. $(p + g)rs = prs + grs$
6. $w + m$ is a unique real number	22. If $x + 1 = y$, then $y = x + 1$
7. $xy = xy$	23. $5m = 5m$
8. $(xy)z = x(yz)$	24. $bc + cd = cd + bc$
9. rs is a unique real number	25. $(9rs)(0) = 0$
10. $162 = x$ implies $x = 162$	26. $4x + 7y$ is a unique real number
11. $5y + 4x = 4x + 5y$	27. $g + (f + e) = (g + f) + e$
12. $0 \cdot b^5 = 0$	28. $(1)(13c^7) = 13c^7$
13. $a + (y + 1) = (a + y) + 1$	29. $3w = 5x$ implies $3wy = 5xy$
14. $a^3 + (-a^3) = 0$	30. $\dfrac{a}{c} \cdot \dfrac{c}{a} = 1$
15. If $10x = 7y$ and $7y = 2b$, then $10x = 2b$	31. $(6r)s = 6(rs)$
16. $\dfrac{5}{3} \cdot \dfrac{3}{5} = 1$	32. $0 = -(5ad) + 5ad$

MAVA Math: Enhanced Skills Copyright © 2015 Marla Weiss

Proportions 1

Solve by simplifying first.

1. $\dfrac{45}{x} = \dfrac{35}{63}$

2. $\dfrac{12}{x} = \dfrac{30}{35}$

3. $\dfrac{15}{35} = \dfrac{21}{x}$

4. $\dfrac{32}{x} = \dfrac{24}{21}$

5. $\dfrac{33}{x} = \dfrac{55}{60}$

6. $\dfrac{34}{x} = \dfrac{51}{15}$

7. $\dfrac{x}{28} = \dfrac{55}{35}$

8. $\dfrac{x}{18} = \dfrac{51}{34}$

9. $\dfrac{32}{x} = \dfrac{52}{39}$

10. $\dfrac{18}{x} = \dfrac{27}{24}$

11. $\dfrac{x}{35} = \dfrac{88}{40}$

12. $\dfrac{16}{x} = \dfrac{20}{35}$

13. $\dfrac{33}{77} = \dfrac{12}{x}$

14. $\dfrac{x}{10} = \dfrac{56}{35}$

15. $\dfrac{x}{15} = \dfrac{26}{65}$

16. $\dfrac{22}{x} = \dfrac{55}{15}$

17. $\dfrac{x}{36} = \dfrac{95}{30}$

18. $\dfrac{x}{45} = \dfrac{78}{54}$

19. $\dfrac{9}{21} = \dfrac{x}{49}$

20. $\dfrac{14}{20} = \dfrac{35}{x}$

21. $\dfrac{x}{65} = \dfrac{64}{80}$

22. $\dfrac{51}{x} = \dfrac{66}{44}$

23. $\dfrac{35}{x} = \dfrac{25}{45}$

24. $\dfrac{66}{x} = \dfrac{72}{84}$

25. $\dfrac{49}{77} = \dfrac{56}{x}$

26. $\dfrac{27}{x} = \dfrac{48}{80}$

27. $\dfrac{36}{x} = \dfrac{81}{99}$

28. $\dfrac{14}{x} = \dfrac{34}{85}$

29. $\dfrac{x}{65} = \dfrac{57}{95}$

30. $\dfrac{x}{48} = \dfrac{77}{66}$

31. $\dfrac{55}{x} = \dfrac{99}{72}$

32. $\dfrac{36}{x} = \dfrac{98}{49}$

33. $\dfrac{x}{28} = \dfrac{54}{36}$

Proportions 2

Solve for x by cross multiplying. Variables are nonzero.

1. $\dfrac{6}{x} = \dfrac{5}{9}$

2. $\dfrac{5}{x} = \dfrac{8}{15}$

3. $\dfrac{7}{3} = \dfrac{5}{x}$

4. $\dfrac{2}{13} = \dfrac{11}{x}$

5. $\dfrac{6}{5} = \dfrac{7}{x}$

6. $\dfrac{5}{x} = \dfrac{9}{4}$

7. $\dfrac{3}{4} = \dfrac{2}{x}$

8. $8 : x = 5 : 12$

9. $\dfrac{h}{e} = \dfrac{c}{x}$

10. $\dfrac{5}{x} = \dfrac{15}{4}$

11. $\dfrac{6}{7} = \dfrac{9}{x}$

12. $\dfrac{6}{x} = \dfrac{10}{9}$

13. $\dfrac{11}{x} = \dfrac{33}{7}$

14. $\dfrac{9}{2x} = \dfrac{5}{4}$

15. $\dfrac{5a}{x} = \dfrac{2b}{y}$

16. $45 : 11 = 10 : x$

17. $\dfrac{9}{x} = \dfrac{13}{11}$

18. $\dfrac{3x}{17} = \dfrac{5}{2}$

19. $\dfrac{7}{5} = \dfrac{12}{x}$

20. $\dfrac{7}{x} = \dfrac{9a}{7}$

21. $\dfrac{12}{x} = \dfrac{18}{5}$

22. $\dfrac{6k}{m} = \dfrac{2a}{x}$

23. $\dfrac{9}{8} = \dfrac{x}{11}$

24. $3a : 7 = 8a : ex$

Proportions 3

Write and solve a proportion, assuming a constant rate.

1.	If a store needs 9 registers open for 150 customers, how many registers are needed for 200 customers?	7.	When a four-foot child casts a 7-foot shadow, find the height of a building in feet with a 168-foot shadow.
2.	If the tax on a $7200 item is $504, find the tax on an item costing $8000.	8.	If 12 sheet cakes serve 26 people, how many people do 54 identical sheet cakes serve?
3.	If 8 workers load 12 trucks, how many workers are needed to load 21 trucks?	9.	A recipe needs 2 cups sugar for every 3 cups flour. How many cups of sugar are needed for 10 cups of flour?
4.	Find the real estate tax on a $80,000 home if the tax on a $110,000 home is $1320.	10.	Find the restaurant-calculated tip in dollars on a $500 bill if a $650 bill at the same restaurant had a $104 tip.
5.	224 ounces of punch are needed to serve 35 people. How many ounces are needed to serve 50 people?	11.	The weight of 64 feet of rope is 20 pounds. What is the weight in pounds of 80 feet of the same rope?
6.	A photograph 5 by 8 inches is enlarged so that one dimension is 7 inches. Find the other new dimension.	12.	If a 13.5-foot flagpole casts a 19.5-foot shadow, what is the shadow of a nearby 54-foot tree at the same time?

Proportions 4

Solve, using the two-column method.

1. Picking 4 bushels of peaches in 25 minutes, how many bushels can be picked in 5 hours?

6. 14 pounds of chips serve 64 people. How many ounces are needed to serve 14 people?

2. If 7 inches of material costs $17.50, find the cost of 2 feet of the same material.

7. If a map uses a scale of 1 cm to 150 km, find the actual distance in km if the map distance is 45 mm.

3. Typing 250 words in 20 minutes, how many hours are needed to type a 7500 word paper?

8. Typing 50 words per 40 seconds, how many words can be typed in 12 minutes?

4. A recipe requires 1.25 cups of sugar for 60 cookies. How many cups are needed for 9 dozen?

9. A photograph 6 by 9 inches is enlarged so that the lesser dimension is 4 feet. Find the other dimension in feet.

5. If a person 79.2 inches tall casts a shadow 9 feet long, find the height of a tree in feet casting a 60-foot long shadow at the same time and place.

10. The weight of 63 feet of rope is 24 pounds. What is the weight in pounds of 56 yards of the same rope?

Proportions 5

Find the number.

1. A number added to 15, then divided by 11 is the same as 14 less than the number, then divided by 12.

2. The quotient of 5 more than a number and 15 equals the quotient of 3 less than the number and 13.

3. Half the sum of 22 and a number equals one-fifth the sum of 16 and the number.

4. The sum of triple a number and 7, then divided by 4, equals 7 less than twice the number, then divided by 5.

5. The quotient of 3 less than 4 times a number and 16 equals the quotient of 8 less than the number and 6.

6. Half the sum of 25 and a number equals one-sixth the sum of 5 and twice the number.

7. A number subtracted from 8, then divided by 10, is the same as 2 less than the number, then divided by 20.

8. Half the sum of 8 and triple a number equals one-third the sum of 7 and twice the number.

9. The quotient of 7 more than a number and 30 equals the quotient of 3 less than twice the number and 50.

10. A number added to 50, then divided by 20, is the same as 16 less than twice the number, then divided by 11.

11. One-third the sum of 17 and a number equals one-sixth the sum of 16 and five times the number.

12. The quotient of 9 more than four times a number and 10 equals the quotient of 4 more than the number and 20.

Proportions 6

Answer as indicated assuming a constant rate. Radii and diameters are in units.	Find the positive geometric mean.
1. If a cookie with diameter 12 serves 4 people, a cookie with diameter 18 serves how many?	9. between 4 and 25
2. A pizza with radius 15 serves how many people if a pizza with radius 10 serves 4 people?	10. between 8 and 50
3. If a silver disk with diameter 40 makes 9 rings, a disk with radius 40 makes how many rings?	11. between 3 and 48
4. A cake with circumference 48π serves 32 people. A cake with circumference 36π serves how many people?	12. between 18 and 50
5. A pie with radius 8 serves how many people if a pie with radius 16 serves 20 people?	13. between 5 and 45
6. A pizza with diameter 20 serves how many people if a pizza with radius 2 serves 5 people?	14. between 11 and 44
7. If a cookie with diameter 6 serves 3 people, a cookie with diameter 24 serves how many?	15. between 6 and 54
8. A cake with circumference 10π serves 7 people. A cake with circumference 30π serves how many people?	16. between 16 and 36

Proportions 7

	Answer as indicated, assuming inverse proportionality.		*Answer as indicated, assuming direct proportionality.*
1.	If 24 people can do a job in 5 days, then how many people are needed to do the job in 1 day?	7.	If the sales tax on an item costing x dollars is y dollars, find the tax on an item costing z dollars.
2.	If the intensity of a light measures 25 at a distance of 3 units from the light, then find the intensity of the light at a distance of 15 units.	8.	How many words can be keyed in s seconds at a rate of w words in m minutes?
3.	If 60 bees need 6 hours to produce a quantity of honey, how many bees are needed to produce the same amount of honey in 5 hours?	9.	The weight of f feet of rope is p pounds. What is the weight in pounds of y yards of the same rope?
4.	If 6 cats can catch 10 mice in 15 days, how many days would be needed for 10 cats to catch the same number of mice?	10.	A recipe requires c cups of sugar for k cookies. How many cups of sugar are needed for d dozen cookies?
5.	If manufacturing 100 items costs $520, then find the cost of manufacturing 325 of the same items.	11.	If c pounds of chips serve p people, how many ounces of chips are needed to serve q people?
6.	If rectangular areas of land have a fixed price per square yard, then an area 32 by 45 yards is equivalent to an area 72 by how many yards?	12.	Picking b bushels of peaches in m minutes, how many bushels can be picked in h hours ?

Proportions 8

Answer as indicated.

1.	X varies directly as Y. X = 60 when Y = 5. Find X when Y = 11.	8.	P varies inversely as Q. P = 75 when Q = 20. Find Q when P = 60.
2.	A varies jointly as B and C. A = 300 when B = 5 and C = 3. Find A when B = 10 and C = 12.	9.	X varies directly as Y and inversely as Z. X = 231 when Y = 22 and Z = 4. Find Z when X = 35 and Y = 10.
3.	M varies inversely as N. M = 40 when N = 12. Find N when M = 24.	10.	M varies directly as the square root of N. M = 220 when N = 400. Find N when M = 165.
4.	M varies inversely as the cube of N. M = 8 when N = 6. Find N when M = 27.	11.	X varies jointly as Y and Z. X = 450 when Y = 10 and Z = 15. Find Z when X = 540 and Y = 9.
5.	R varies directly as the square of P and inversely as Q. R = 4 when P = 5 and Q = 3. Find R when P = 10 and Q = 15.	12.	A varies jointly as B, C, and D. A = 360 when B = 5, C = 9, and D = 7. Find A when B = 14, C = 3, and D = 2.
6.	X varies directly as Y. X = 57 when Y = 6. Find X when Y = 22.	13.	T varies directly as the cube of S and inversely as R. T = 63 when R = 4 and S = 3. Find R when S = 3 and T = 2.
7.	Y varies directly as X. X = 60 when Y = 5. Find X when Y = 11.	14.	R varies jointly as P and Q and inversely as the square of T. R = 12 when P = 7, Q = 6, and T = 14. Find R when T = 2, P = 3 and Q = 5.

Pyramids 1

Find the volume of the pyramid in cubic units.

1. square base, edge 12 altitude 8	9. trapezoid base, sides 5, 5, 5, and 11 altitude 15
2. equilateral triangle base, edge 12 altitude 16	10. regular hexagon base, edge 10 altitude 6
3. rectangle base, length 12, width 8 altitude 11	11. square base, edge 15 altitude 11
4. square base, diagonal 18 altitude 5	12. triangle base, edges 5, 5, and 6 altitude 13
5. equilateral triangle base, edge 6 altitude 7	13. octagon base, area 51 altitude 10
6. rectangle base, length 15, width 14 altitude 24	14. greatest pyramid that fits in a box with base 5 by 5 and height 12
7. regular hexagon base, edge 6 altitude 20	15. triangle base, edges 25, 25, and 14 altitude 9
8. pentagon base, area 48 altitude 11	16. greatest pyramid that fits in a box with base 6 by 7 and height 11

Pyramids 2

Find the total surface area of the pyramid in square units.

1.   square base, edge 16
     lateral edge 10
	9. square base, edge 6
	     altitude 4

2.   rectangle base, length 14, width 40
     lateral edge 25
	10. rectangle base, length 10, width 18
	      altitude 12

3.   equilateral triangle base, edge 16
     lateral edge 17
	11. square base, edge 24
	      altitude 35

4.   square base, edge 24
     lateral edge 13
	12. square base, edge 10
	      lateral edge 13

5.   rectangle base, length 80, width 26
     lateral edge 85
	13. rectangle base, length 60, width 18
	      altitude 40

6.   equilateral triangle base, edge 40
     lateral edge 29
	14. rectangle base, length 14, width 36
	      altitude 24

7.   square base, edge 14
     lateral edge 25
	15. square base, edge 32
	      altitude 30

8.   rectangular base, length 14, width 30
     lateral edge 25
	16. rectangle base, length 22, width 50
	      altitude 60

Pyramids 3

Find the altitude of the pyramid in units.	Find the volume of the pyramid in cubic units.

1. square base, edge 10
    volume 1000
	9. square base
	    altitude 24, slant height 26

2. rectangle base, length 12, width 8
    volume 352
	10. square base
	     altitude 12, slant height 15

3. rhombus base with diagonals 9 and 10
    volume 225
	11. square base, area 40
	     slant height 8

4. trapezoid base, edges 13, 13, 13, 23
    volume 504
	12. square base
	     altitude 13, slant height 22

5. trapezoid base, edges 10, 10, 10, 26
    volume 324
	13. square base, area 72
	     slant height 9

6. rhombus base with diagonals 8 and 9
    volume 156
	14. square base
	     altitude 30, slant height 34

7. square base, edge 12
    volume 624
	15. square base
	     altitude 24, slant height 25

8. rectangular base, length 9, width 15
    volume 765
	16. square base
	     altitude 60, slant height 61

Pyramids 4

Find the volume in cubic units of the pyramid frustrum.

1. square base with edge 3, upper edge 1, frustrum height 6

2. triangular base with area 64, pyramid altitude 24, cut 9 up from base

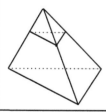

3. square base with edge 30, pyramid altitude 25, cut 15 down from vertex

4. triangular base with area 49, pyramid altitude 21, cut 6 up from base

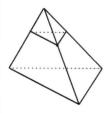

5. triangular base with area 100, pyramid altitude 30, frustrum altitude 9

6. square base with edge 24, pyramid altitude 10, cut halfway up

7. square base with edge 15, upper edge 3, frustrum height 8

8. square base with edge 8, upper edge 6, frustrum height 3

9. triangular base with area 27, pyramid altitude 15, cut 5 up from base

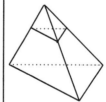

10. square base with edge 12, pyramid altitude 20, cut 5 up from base

Pythagorean Theorem 1

Determine whether three numbers form the sides of a right triangle (YES/NO). Give a reason.

1. 9, 40, 41	14. 24, 70, 72
2. 9, 12, 16	15. 20, 21, 28
3. 24, 32, 45	16. 1, 2.4, 2.6
4. 4, 5, 6	17. 10, 10.5, 14.5
5. 2, 3, $\sqrt{13}$	18. 4, 5, $\sqrt{43}$
6. 14, 48, 50	19. 6, 7, $\sqrt{85}$
7. 16, 30, 34	20. 2.5, 6, 6.5
8. 4, 4, $4\sqrt{2}$	21. 5, 9, $\sqrt{106}$
9. 11, 50, 51	22. 24, 32, 40
10. 7, 7, $7\sqrt{2}$	23. 3, 7, $\sqrt{59}$
11. 40, 42, 48	24. 2, 5, $\sqrt{29}$
12. 1, 3, $\sqrt{10}$	25. 5, 10, $5\sqrt{5}$
13. 1, 1, $\sqrt{2}$	26. $\sqrt{5}$, 3, 4

Pythagorean Theorem 2

Find the missing side (in ascending order) of the Pythagorean triplet by mental math.

1. 3, 4, _____	17. 14, 48, _____	33. _____ , 168, 170
2. 7, 24 _____	18. 2, 2, _____	34. 33, 180, _____
3. 8, _____ , 17	19. 40, 42, _____	35. 90, _____ , 410
4. 5, 12, _____	20. 5, _____ , 10	36. 14, 14, _____
5. 5, 5, _____	21. 1, _____ , 2	37. 12, _____ , 20
6. 9, 40, _____	22. 22, _____ , 122	38. 32, 60, _____
7. 11, 60, _____	23. _____ , 36, 39	39. 77, 264, _____
8. 12, _____ , 37	24. 80, 150, _____	40. 18, _____ , 30
9. 14, _____ , 50	25. 33, _____ , 55	41. 24, _____ , 74
10. 10, _____ , 20	26. 16, _____ , 32	42. 9, _____ , 18
11. 18, 80, _____	27. 3, 3, _____	43. 25, _____ , 65
12. 10, _____ , 26	28. 13, 84, _____	44. 132, 385, _____
13. _____ , 180, 181	29. _____ , 12, 15	45. 24, _____ , 51
14. 20, _____ , 29	30. 44, 240, _____	46. 30, 40, _____
15. 6, 8, _____	31. 16, 30, _____	47. 3, _____ , 6
16. 33, 33, _____	32. 21, 28, _____	48. 36, 105, _____

Pythagorean Theorem 3

Draw a picture and calculate.

1. A tree broke 36 feet above the ground. The attached top rests on the ground 27 feet from the bottom of the tree. How many feet tall was the tree?	6. A clock has a minute hand 16 cm long and an hour hand 12 cm long. How many cm apart are the ends of the hands at 3:00?
2. 30 steps east, 10 steps north, 10 steps east, and finally 20 steps north is how many direct steps from the starting point?	7. A car went 12 miles west, 7 miles south, and then 12 miles west. How many miles was the car from its starting point?
3. Jo and Bo left at the same place and time. Jo walked south at 3 mph, while Bo ran east at 4 mph. How far apart were they 2 hours later?	8. Jan and Hal left work at 6:00 PM. Jan drove north at 35 mph, while Hal sped west at 84 mph. How far apart were they at 6:30 PM?
4. If a person walks 5 miles east, 5 miles south, and then 7 miles east, how many miles would he have saved by traveling in a straight line?	9. 30 steps east, 10 steps north, 6 steps west, and 3 steps south is how many steps from the starting point?
5. The front of a boat is tied by a 20-foot rope to a point on a pier that is 12 feet above the height of the boat. If 5 feet of the rope are pulled in, how many feet forward will the boat move?	10. Two boats leave from the same dock at 11:00AM. One goes north at 60mph, while the other travels east at 80mph. At what time will they be 150 miles part?

Pythagorean Theorem 4

Find the distance between the pair of points.

1. (6, –6) and (–2, 9)

2. (5, –8) and (–5, 16)

3. (–8, –21) and (2, 3)

4. (–8, –2) and (16, 5)

5. (9, –11) and (–15, –4)

6. (10, 3) and (–14, –4)

7. (3, –8) and (–2, 4)

8. (–10, –11) and (6, 19)

9. (–1, –11) and (7, –5)

10. (4, –8) and (–4, 7)

11. (6, 8) and (–18, –2)

12. (7, 0) and (0, 7)

13. (5, 1) and (–4, –1)

14. (2, 5) and (5, –1)

15. (–4, –2) and (0, 5)

16. (3, 4) and (–1, –1)

17. (7, 2) and (5, 9)

18. (1, 2) and (6, 5)

19. (5, 1) and (4, –8)

20. (–2, 1) and (–1, 2)

21. (3, 10) and (–1, 6)

22. (11, 10) and (7, 4)

Pythagorean Theorem 5

Given measures for 3 sides, indicate whether the triangle is acute or obtuse.

1. 6, 12, 13	12. 5, 12, 14	23. 2, 4, 5
2. 4, 5, 7	13. 4, 5, 6	24. 3, 6, 8
3. 8, 11, 13	14. 2, 3, 4	25. 4, 9, 10
4. 2, 4, 5	15. 5, 7, 11	26. 10, 15, 18
5. 10, 11, 15	16. 7, 8, 9	27. 5, 7, 9
6. 5, 10, 12	17. 6, 10, 15	28. 4, 11, 13
7. 7, 9, 12	18. 8, 10, 11	29. 2, 6, 7
8. 5, 6, 8	19. 10, 11, 12	30. 12, 15, 19
9. 5, 8, 11	20. 5, 6, 7	31. 9, 10, 15
10. 6, 7, 8	21. 3, 9, 10	32. 8, 9, 10
11. 10, 13, 15	22. 10, 20, 25	33. 6, 8, 11

Pythagorean Theorem 6

Answer as indicated.

1. Find the sum of the perimeter and area of the square if the legs of the triangle are 6 and 8.

6. Find AB.

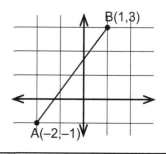

2. Find AB if the 3 congruent squares each have area 9.

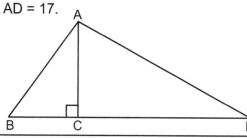

7. Find FH if EG = 42, EF = 26, and FG = 40.

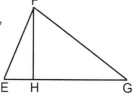

3. Find AC if BC = 18, CD = 14, and AD = 40.

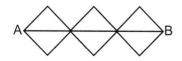

8. The areas of the 3 squares are 4, 9, and 16. Find the length of the diagonal line shown.

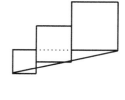

4. Find AC if AB = 10, BD = 21, and AD = 17.

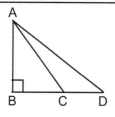

9. Find the perimeter of the parallelogram.

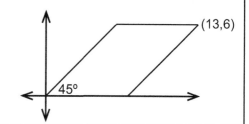

5. Find the sum of all of the diagonals of the three squares if the largest square has area 16.

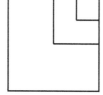

10. ABCD is a 16 by 24 rectangle. F is a midpoint. BE = 7. Find the perimeter of △AEF.

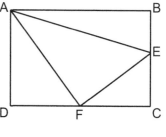

Pythagorean Theorem 7

Find the area of the rectangle in square units.	*Find the perimeter of the isosceles right triangle in units.*
1. width 2, diagonal 5	10. leg 6
2. length 10, diagonal 14	11. hypotenuse $\sqrt{6}$
3. width 6, diagonal 25	12. hypotenuse 6
4. length 18, diagonal 20	13. leg $\sqrt{6}$
5. length 14, diagonal 24	14. hypotenuse 10
6. length 30, diagonal 33	15. leg 10
7. length 15, diagonal 35	16. leg $\sqrt{10}$
8. length 16, diagonal 44	17. hypotenuse $\sqrt{10}$
9. width 8, diagonal 9	18. hypotenuse $\sqrt{2}$

Pythagorean Theorem 8

Find the number of right triangles with whole number sides that have one side as given.

1. 5	8. 16	15. 40
2. 8	9. 20	16. 48
3. 9	10. 24	17. 60
4. 10	11. 25	18. 100
5. 12	12. 30	19. 144
6. 14	13. 35	20. 160
7. 15	14. 36	21. 200

Radicals 1

Simplify the radical.

1. $\sqrt{1}$	17. $\sqrt{(47)^2}$	33. $\sqrt{\dfrac{400}{361}}$
2. $\sqrt{49}$	18. $\sqrt{64}$	34. $\sqrt{121,000,000}$
3. $\sqrt{256}$	19. $\sqrt{900}$	35. $\sqrt{324}$
4. $\sqrt{4900}$	20. $\sqrt{(91)^2}$	36. $\sqrt{961}$
5. $\sqrt{(39)^2}$	21. $\sqrt{\dfrac{16}{49}}$	37. $\sqrt{(111)^2}$
6. $\sqrt{225}$	22. $\sqrt{441}$	38. $\sqrt{\dfrac{4900}{22,500}}$
7. $\sqrt{\dfrac{81}{25}}$	23. $\sqrt{486^2}$	39. $\sqrt{8100}$
8. $\sqrt{6400}$	24. $\sqrt{1600}$	40. $\sqrt{151^2}$
9. $\sqrt{360,000}$	25. $\sqrt{28,900}$	41. $\sqrt{1,440,000}$
10. $\sqrt{121}$	26. $\sqrt{625}$	42. $\sqrt{\dfrac{324}{289}}$
11. $\sqrt{10,000}$	27. $\sqrt{32,400}$	43. $\sqrt{160,000}$
12. $\sqrt{12,100}$	28. $\sqrt{617^2}$	44. $\sqrt{1,000,000}$
13. $\sqrt{\dfrac{1}{9}}$	29. $\sqrt{\dfrac{121}{625}}$	45. $\sqrt{\dfrac{169}{289}}$
14. $\sqrt{\dfrac{90}{40}}$	30. $\sqrt{361}$	46. $\sqrt{2,560,000}$
15. $\sqrt{144}$	31. $\sqrt{\dfrac{49}{64}}$	47. $\sqrt{810,000}$
16. $\sqrt{400}$	32. $\sqrt{\dfrac{441}{961}}$	48. $\sqrt{\dfrac{361}{196}}$

Radicals 2

Simplify the radical.

1. $\sqrt{18}$	12. $\sqrt{108}$	23. $\sqrt{245}$
2. $\sqrt{27}$	13. $\sqrt{125}$	24. $\sqrt{275}$
3. $\sqrt{40}$	14. $\sqrt{128}$	25. $\sqrt{320}$
4. $\sqrt{48}$	15. $\sqrt{147}$	26. $\sqrt{360}$
5. $\sqrt{50}$	16. $\sqrt{150}$	27. $\sqrt{363}$
6. $\sqrt{54}$	17. $\sqrt{162}$	28. $\sqrt{432}$
7. $\sqrt{72}$	18. $\sqrt{175}$	29. $\sqrt{448}$
8. $\sqrt{75}$	19. $\sqrt{180}$	30. $\sqrt{567}$
9. $\sqrt{80}$	20. $\sqrt{192}$	31. $\sqrt{675}$
10. $\sqrt{90}$	21. $\sqrt{242}$	32. $\sqrt{800}$
11. $\sqrt{98}$	22. $\sqrt{243}$	33. $\sqrt{847}$

Radicals 3

Simplify and multiply.

1. $\sqrt{10} \cdot \sqrt{15} \cdot \sqrt{6}$	12. $\sqrt{18} \cdot \sqrt{14} \cdot \sqrt{21}$	23. $5\sqrt{35} \cdot 6\sqrt{15}$
2. $\sqrt{26} \cdot \sqrt{39} \cdot \sqrt{6}$	13. $\sqrt{20} \cdot \sqrt{21} \cdot \sqrt{35}$	24. $2\sqrt{8} \cdot 4\sqrt{14}$
3. $\sqrt{35} \cdot \sqrt{21} \cdot \sqrt{15}$	14. $\sqrt{36} \cdot \sqrt{35} \cdot \sqrt{28}$	25. $3\sqrt{6} \cdot 2\sqrt{10}$
4. $\sqrt{34} \cdot \sqrt{51} \cdot \sqrt{6}$	15. $\sqrt{12} \cdot \sqrt{33} \cdot \sqrt{55}$	26. $4\sqrt{24} \cdot 5\sqrt{20}$
5. $\sqrt{98} \cdot \sqrt{25} \cdot \sqrt{32}$	16. $\sqrt{18} \cdot \sqrt{10} \cdot \sqrt{6}$	27. $5\sqrt{27} \cdot 2\sqrt{18}$
6. $\sqrt{36} \cdot \sqrt{28} \cdot \sqrt{7}$	17. $\sqrt{72} \cdot \sqrt{99} \cdot \sqrt{44}$	28. $5\sqrt{39} \cdot 2\sqrt{26}$
7. $\sqrt{21} \cdot \sqrt{14} \cdot \sqrt{6}$	18. $\sqrt{75} \cdot \sqrt{48} \cdot \sqrt{18}$	29. $3\sqrt{60} \cdot 3\sqrt{45}$
8. $\sqrt{27} \cdot \sqrt{15} \cdot \sqrt{5}$	19. $\sqrt{54} \cdot \sqrt{21} \cdot \sqrt{3}$	30. $3\sqrt{55} \cdot 5\sqrt{33}$
9. $\sqrt{54} \cdot \sqrt{24} \cdot \sqrt{64}$	20. $\sqrt{40} \cdot \sqrt{50} \cdot \sqrt{60}$	31. $3\sqrt{22} \cdot 2\sqrt{44}$
10. $\sqrt{28} \cdot \sqrt{80} \cdot \sqrt{35}$	21. $\sqrt{24} \cdot \sqrt{25} \cdot \sqrt{27}$	32. $7\sqrt{48} \cdot 8\sqrt{12}$
11. $\sqrt{48} \cdot \sqrt{63} \cdot \sqrt{21}$	22. $\sqrt{35} \cdot \sqrt{14} \cdot \sqrt{81}$	33. $9\sqrt{42} \cdot 4\sqrt{28}$

Radicals 4

Simplify and add or subtract.

1. $5\sqrt{28} - 2\sqrt{63}$	9. $5\sqrt{12} - 2\sqrt{27}$	17. $7\sqrt{20} + 6\sqrt{45}$
2. $2\sqrt{48} + 4\sqrt{147}$	10. $6\sqrt{50} - 2\sqrt{8}$	18. $\sqrt{128} + 2\sqrt{98}$
3. $6\sqrt{18} + 5\sqrt{72}$	11. $5\sqrt{20} - \sqrt{125}$	19. $\sqrt{200} - 2\sqrt{32}$
4. $\sqrt{175} + 3\sqrt{63}$	12. $\sqrt{150} + 2\sqrt{54}$	20. $\sqrt{176} + 3\sqrt{44}$
5. $5\sqrt{75} - 2\sqrt{300}$	13. $5\sqrt{99} - \sqrt{275}$	21. $\sqrt{500} - 2\sqrt{80}$
6. $\sqrt{288} + 2\sqrt{800}$	14. $\sqrt{147} + 2\sqrt{108}$	22. $\sqrt{171} + 3\sqrt{76}$
7. $\sqrt{720} + 5\sqrt{245}$	15. $\sqrt{363} + 2\sqrt{75}$	23. $6\sqrt{52} - 2\sqrt{117}$
8. $\sqrt{252} + 2\sqrt{448}$	16. $\sqrt{600} - 4\sqrt{24}$	24. $\sqrt{112} + 2\sqrt{343}$

Radicals 5

Round the radical as specified. Use a calculator.	*Simplify the radical. Use mental math.*
1. $\sqrt{37}$ to the nearest thousandth	14. $\sqrt{2.56}$
2. $\sqrt{59}$ to the nearest tenth	15. $\sqrt{0.000064}$
3. $\sqrt{91}$ to the nearest hundredth	16. $\sqrt{6.25}$
4. $\sqrt{67}$ to the nearest ten-thousandth	17. $\sqrt{0.0049}$
5. $\sqrt{11}$ to the nearest thousandth	18. $\sqrt{.81}$
6. $\sqrt{29}$ to the nearest hundredth	19. $\sqrt{0.0361}$
7. $\sqrt{63}$ to the nearest tenth	20. $\sqrt{.36}$
8. $\sqrt{95}$ to the nearest ten-thousandth	21. $\sqrt{0.0256}$
9. $\sqrt{55}$ to the nearest ten-thousandth	22. $\sqrt{(1.8)^2}$
10. $\sqrt{97}$ to the nearest hundredth	23. $\sqrt{0.000016}$
11. $\sqrt{77}$ to the nearest thousandth	24. $\sqrt{1.21}$
12. $\sqrt{31}$ to the nearest hundredth	25. $\sqrt{0.0324}$
13. $\sqrt{23}$ to the nearest thousandth	26. $\sqrt{1.69}$

Radicals 6

Rationalize the denominator. Simplify.

1. $\dfrac{20}{\sqrt{5}}$	9. $\dfrac{20}{\sqrt{5}+3}$	17. $\dfrac{57}{2\sqrt{5}+1}$
2. $\dfrac{75}{\sqrt{3}}$	10. $\dfrac{25}{\sqrt{6}+1}$	18. $\dfrac{55}{3\sqrt{4}-5}$
3. $\dfrac{14}{\sqrt{2}}$	11. $\dfrac{12}{\sqrt{7}-2}$	19. $\dfrac{42}{5\sqrt{2}+6}$
4. $\dfrac{63}{\sqrt{7}}$	12. $\dfrac{42}{\sqrt{3}-3}$	20. $\dfrac{-13}{4\sqrt{3}+7}$
5. $\dfrac{66}{\sqrt{11}}$	13. $\dfrac{16}{\sqrt{9}+5}$	21. $\dfrac{60}{2\sqrt{6}-3}$
6. $\dfrac{42}{\sqrt{6}}$	14. $\dfrac{18}{\sqrt{6}-3}$	22. $\dfrac{58}{3\sqrt{5}+4}$
7. $\dfrac{80}{\sqrt{10}}$	15. $\dfrac{28}{\sqrt{2}-4}$	23. $\dfrac{45}{4\sqrt{4}+7}$
8. $\dfrac{60}{\sqrt{15}}$	16. $\dfrac{56}{\sqrt{8}-1}$	24. $\dfrac{44}{5\sqrt{3}-8}$

Ratios 1

Write each ratio first as identifying letters and then as a simplified fraction.

1. number of letters to vowels in the word SCHOOLHOUSE

2. number of vowels to consonants in the word SCHOOLHOUSE

3. number of leap years to years from 2000 through 2009

4. number of months of the year to months with 31 days

5. number of primes to composites from 1 through 30

6. number of prime digits to composite digits in 7,535,197,228

7. number of percent signs to dollar signs in the string ??$#@$%*?%#??+@$?&

8. number of blues to greens in the list Navy, Grass, Sky, Forest, Lime

9. number of even digits to odd digits in 5,347,190,943

10. number of vowels (without Y) to letters in the alphabet

11. number of face cards to total cards in a standard deck

12. number of letters to vowels in the word BOOKFAIR

13. number of consonants to vowels to in the word WORKSHOP

14. number of leap years to years from 1900 through 1999

15. number of months with 30 days to months of the year

16. number of primes to total numbers from 1 through 20

17. number of prime digits to composite digits in 6,035,831,396,491

18. number of asterisks to percents in the string *%*#&*%*?*#%&+%*$

19. number of tans to pinks in the list Beige, Hot, Magenta, Taupe

20. number of odd digits to even digits in 7,815,240,968

21. number of vowels (with Y) to letters in the alphabet

22. number of face cards to even cards (with Q) in a standard deck

Ratios 2

Answer as indicated.

1.	The sum of two whole numbers is 36. If the numbers are in the ratio 5:4, find their product.	8.	In a class if the ratio of the number of girls to boys is 3:5, find the ratio of the number of boys to students.
2.	If $27,000 were divided among three people in the ratio 2:3:5, find the value of the greatest share.	9.	A recipe has the ratio of cups of flour to sugar to butter as 4:2:1. If the recipe is doubled, find the new ratio of the three ingredients.
3.	If $30,000 were divided among three people in the ratio 3:5:7, find the value of the least share.	10.	If the ratio of a to b is 5 to 8 and the ratio of b to c is 4 to 3, find the ratio of a to c.
4.	The sum of two whole numbers is 60. If the numbers are in the ratio 7:5, find their product.	11.	In a class if the ratio of the number of students to boys is 11:5, find the ratio of the number of boys to girls.
5.	Three whole numbers have a sum of 80 and are in the ratio 2:5:9. Find the least of the numbers.	12.	The ratio of x to 50 equals the ratio of 18 to 30. Find x.
6.	If 600 items were divided four ways in the ratio 3:2:8:2, find the median value.	13.	The ratio of Jo's age to Tom's age is 2:3. The ratio of Bo's age to Tom's age is 8:7. Find the ratio of Bo's age to Jo's age.
7.	Three whole numbers have a sum of 72 and are in the ratio 1:5:6. Find the greatest of the numbers.	14.	If the ratio of c to d is 7 to 9 and the ratio of d to e is 27 to 8, find the ratio of e to c.

Ratios 3

Find the unit rate.	Find the cost for the number specified, assuming a constant rate.
1. $55 for 25 candles	12. 7 if 3 cost $111
2. $70 for 42 pens	13. 9 if 4 cost $236
3. $165 for 132 pencils	14. 11 if 5 cost $425
4. $130 for 78 tags	15. 8 if 6 cost $522
5. $105 for 90 buttons	16. 3 if 11 cost $572
6. 572 apples for 44 crates	17. 3 if 10 cost $13.20
7. 204 people for 51 cars	18. 13 if 8 cost $16.40
8. 336 students for 21 teachers	19. 7 if 15 cost $360
9. 88 pies for 132 people	20. 10 if 6 cost $30.90
10. 600 pounds for 25 boxes	21. 12 if 7 cost $21.28
11. 209 pounds for 22 cartons	22. 5 if 12 cost $72.24

Ratios 4

Find the better buy using mental math.	*Find the scale or dimensions by mental math.*
1. 2 for $9.60 or 3 for $14.25	12. actual dimensions 605 m by 990 m drawn dimensions 11 cm by 18 cm scale: _____
2. 3 for $13.65 or 4 for $18.24	13. actual dimensions 99 yd by 176 yd drawn dimensions 9 in by 16 in scale: _____
3. 5 for $23.55 or 6 for $28.50	14. actual dimensions 120 yd by 165 yd drawn dimensions 8 in by 11 in scale: _____
4. 11 for $171.60 or 9 for $139.50	15. actual dimensions 98 km by 126 km drawn dimensions 7 cm by 9 cm scale: _____
5. 6 for $5.88 or 7 for $6.93	16. actual dimensions 324 yd by 244 yd scale 1 in = 8 yd drawn dimensions: _____
6. 5 for $178.25 or 2 for $71.34	17. drawn dimensions 5.5 cm by 7 cm scale 1 cm = 26 m actual dimensions: _____
7. 8 for $9.20 or 12 for $13.92	18. actual dimensions 110 ft by 84 ft scale 1 in = 20 ft drawn dimensions: _____
8. 15 for $32.25 or 19 for $39.90	19. drawn dimensions 3.5 cm by 2.1 cm scale 1 cm = 30 m actual dimensions: _____
9. 13 for $97.50 or 11 for $83.60	20. actual dimensions 65 yd by 117 yd drawn dimensions 10 in by 18 in scale: _____
10. 6 for $14.40 or 8 for $19.28	21. drawn dimensions 2.5 cm by 9 cm scale 1 cm = 2.5 m actual dimensions: _____
11. 10 for $29.50 or 9 for 25.20	22. actual dimensions 420 ft by 115.5 ft scale 1 in = 10.5 ft drawn dimensions: _____

Ratios 5

Find the number that can be added to both values of the 1st ratio to get the 2nd ratio.	*A, B, C, D, and E are consecutive points on a line. Find the specified ratio.*
1. 1st ratio 3:7 2nd ratio 3:4	8. AB:BC = 1:2 BC:CD = 2:3 CD:DE = 3:4 Find AC:AE.
2. 1st ratio 2:5 2nd ratio 8:9	9. AB:BC = 1:1 BC:CD = 1:3 CD:DE = 2:5 Find AB:CE.
3. 1st ratio 5:12 2nd ratio 11:12	10. AB:BC = 1:4 BC:CD = 1:2 CD:DE = 1:3 Find AC:BE.
4. 1st ratio 2:9 2nd ratio 4:5	11. AB:BC = 2:3 BC:CD = 3:7 CD:DE = 1:5 Find AD:BE.
5. 1st ratio 2:13 2nd ratio 7:8	12. AC:CD = 3:5 BC:CD = 1:2 CD:DE = 1:3 Find AE:BD.
6. 1st ratio 3:8 2nd ratio 5:6	13. AB:BD = 1:10 BD:DE = 2:3 BC:DE = 2:5 Find CD:AC.
7. 1st ratio 2:9 2nd ratio 6:7	14. AC:CD = 3:2 BC:CD = 1:3 DE:CD = 5:3 Find AB:CE.

Ratios 6

Answer as indicated for the plane geometry ratios.

1. Find the ratio of a side of an equilateral triangle to its perimeter.

2. Given two equilateral triangles, one with side 6 and one with side 20. Find the ratio of the perimeters, larger triangle to smaller.

3. The ratio of a rectangle's length to width is 5:2. If the width is 8, find the area.

4. The measures of three angles of a triangle are in the ratio 2:3:4. Find the measure of the least angle.

5. Find the ratio of a side of a regular hexagon to its perimeter.

6. A rectangle has sides in the ratio 3:5. If the perimeter is 160, find the length of the longer side.

7. The measures of the angles of a quadrilateral are in the ratio 1:2:3:4. Find the sum of the measures of the two greatest angles.

8. The ratio of a rectangle's length to width is 4:3. If the width is 48, find the perimeter.

9. Find the ratio of the area of the hexagon to the area of the 8 by 8 square. The legs of the two isosceles right triangles are 2.

10. Find the ratio of the sides, smaller to larger, of equilateral triangles with perimeters 36 and 90.

11. Find the ratio of Area I to Area II in the rectangle.

12. A rectangle has sides in the ratio 2:9. If the perimeter is 528, find the length of the shorter side.

13. The radii of the smaller circles are 2 and 3. Find the ratio of the areas, smallest circle to largest circle.

14. A rectangle with area 1452 has length and width in the ratio 3:4. Find the diagonal.

Ratios 7

Find the specified ratio.

1. A:B = 5:3
   C:B = 9:7
   Find 2A:B.

2. X:Y = 5:11
   Y:Z = 3:7
   Find (X+Y):(X+Z).

3. A:B = 6:7
   B:C = 4:3
   Find (A + 2B):B.

4. D:O:G = 5:4:9
   G:H:J = 7:6:8
   Find O:J.

5. P:A:N = 35:5:12
   T:O:P = 9:8:15
   Find A:T.

6. A:B:C:D = 3:2:5:7
   C:P:Q:R = 4:7:8:9
   Find R:A.

7. B:I:G = 8:4:3
   B:U:S = 12:5:4
   Find U:G.

8. H:O:T = 7:8:20
   T:I:P = 25:9:21
   Find H:P.

9. S:A:T = 5:9:12
   N:O:T = 6:11:15
   Find A:N.

10. F:U:N = 7:22:5
    S:U:B = 8:33:14
    Find B:F.

11. C:A:N = 11:8:9
    N:O:T= 15:12:13
    Find O:A.

12. S:U:N = 9:10:8
    T:A:N = 12:11:6
    Find T:S.

13. A:B:C = 6:5:14
    H:A:T = 10:7:15
    Find T:B.

14. C:A:K:E = 9:4:6:8
    S:E:N:D = 2:3:5:7
    Find A:D.

15. E:F = 2:9
    D:F = 7:10
    Find 4D:3E.

16. G:H = 9:5
    E:G = 4:11
    Find (E + H):7E.

Ratios 8

Find the ratio of an angle to its complement or supplement as indicated. | *Answer as indicated.*

1. angle to complement if the ratio of the angle to its supplement is 2:7

2. angle to supplement if the ratio of the angle to its complement is 1:8

3. angle to complement if the ratio of the angle to its supplement is 3:7

4. angle to complement if the ratio of the angle to its supplement is 4:5

5. angle to complement if the ratio of the angle to its supplement is 7:8

6. angle to supplement if the ratio of the angle to its complement is 1:9

7. angle to complement if the ratio of the angle to its supplement is 5:7

8. For a cube with edge e, find the ratio of the volume (in cubic units) to the surface area (in square units).

9. The radius of circle A equals the diameter of circle B. Find the ratio of the area of circle A to the area of circle B.

10. For a square with side s, find the ratio of the area (in square units) to the perimeter (in units)?

11. A sphere "just fits" in a cylinder. Find the ratio of the volume of the sphere to the volume of the cylinder.

12. The ratio of the diameters of two circles is 1:4. Find the ratio of their radii.

13. A square and a circle have the same area. Find the ratio of a diagonal of the square to the radius of the circle.

14. Two similar triangles have areas 75 and 108. Find the ratio of their perimeters.

Ratios 9

Find the ratio of the areas as indicated. Inscribed polygons or circles are drawn to midpoints when apparent.

1. inner square to outer square	7. inner triangle to outer equilateral triangle
2. inner square to outer square	8. isosceles right triangle to circle with triangle drawn on diameter
3. circle to square	9. equilateral triangle to circle
4. square to circle	10. regular hexagon to circle
5. inner circle to outer circle	11. smaller circle to larger concentric circle
6. inner square to outer square	12. smaller circle to larger circle

Ratios 10

Find the ratio of the volumes of the space figures that share a base.

1. A cone just fits inside a hemisphere. Find the ratio of the volumes, cone to hemisphere.	5. A sphere just fits in a cube. Find the ratio of the volumes, sphere to cube.
2. A cone just fits in a cylinder. Find the ratio of the volumes, cone to cylinder.	6. A sphere just fits in a cylinder. Find the ratio of the volumes, sphere to cylinder.
3. A cone just fits in a cube. Find the ratio of the volumes, cone to cube.	7. A hemisphere just fits in a rectangular prism. Find the ratio of the volumes, hemisphere to prism.
4. A square pyramid just fits inside a cube. Find the ratio of the volumes, pyramid to cube.	8. A hemisphere just fits in a cylinder. Find the ratio of the volumes, hemisphere to cylinder.

Rectangles 1

Find the area and perimeter of the rectangle by mental math.

	BASE	HEIGHT	AREA	PERIMETER		BASE	HEIGHT	AREA	PERIMETER
1.	36	11			20.	13	13		
2.	14	14			21.	74	11		
3.	5	16			22.	52	10		
4.	51	9			23.	9	25		
5.	3	17			24.	30	4.5		
6.	20	5.5			25.	5	12		
7.	12	1.5			26.	3.5	22		
8.	30	50			27.	40	25		
9.	8	20			28.	11	27		
10.	11	43			29.	8	31		
11.	12	3.5			30.	50	90		
12.	6	18			31.	7	51		
13.	19	3			32.	13	100		
14.	7	21			33.	8	15		
15.	9	13			34.	20	60		
16.	10	16			35.	25	16		
17.	16	4.25			36.	10	6.75		
18.	20	5.75			37.	14	2.25		
19.	11	55			38.	11	38		

Rectangles 2

Answer as indicated.

1. Find the perimeter of a new square formed when an 8 by 8 square has its area reduced by 15 square units.

2. Square A has one-fourth the area of square B. Find the perimeter of square B if the perimeter of square A is 40.

3. Find the maximum area of a rectangle with diagonal 10.

4. A square with area 36 and a rectangle with area 72 have one common congruent side. Find the perimeter of the rectangle.

5. The square is divided into 5 congruent rectangles, each with perimeter 30. Find the perimeter of the square.

6. A rectangle measures 3x + 5 by 6. Find 2 other rectangles with the same area in terms of x.

7. Find the perimeter of a new square formed when a 10 by 10 square has its area reduced by 19 square units.

8. A square with area 20.5 and a rectangle with whole sides have congruent diagonals. Find the rectangle's area and perimeter.

9. Find the sum of the numeric values of the area and perimeter of a square if they are equal.

10. Find the area of a rectangle if the area equals the perimeter and the length is twice the width.

11. A square with diagonal 6 and a rectangle with length 9 have equal areas. Find the diagonal of the rectangle.

12. Three rectangles are shown. Find the perimeter of the greatest. NTS

68   60   75

Rectangles 3

Complete the chart for the rectangle.

	L	W	A	P	D
1.	11			32	
2.		3	27		
3.		10			26
4.	24				25
5.	13			40	
6.	16				34
7.		60			61
8.		35	420		
9.		36		102	
10.		15		54	
11.	12			64	
12.	4		32		
13.	18		36		
14.		13	117		
15.		12	72		
16.	21				29
17.	8			36	
18.		16			20
19.	5			30	

Complete the chart for the square.

	SIDE	AREA	PERIM	DIAG
20.			52	
21.		81		
22.		1		
23.		2		
24.		6		
25.		1.44		
26.		30		
27.				20
28.				12
29.			48	
30.				10
31.			96	
32.		2.25		
33.				8
34.	4			
35.			80	
36.			44	
37.				16
38.		1.69		

Rectangles 4

Find the area in square units algebraically.	*Find the perimeter in units algebraically.*
1. The length of a rectangle is twice its width. Find the area if the perimeter is 120.	7. The length of a rectangle is twice its width. Find the perimeter if the area is 72.
2. The width of a rectangle is 6 more than its length. Find the area if the perimeter is 100.	8. The width of a rectangle is 3 more than its length. Find the perimeter if the area is 340.
3. The length of a rectangle is 4 more than triple the width. If the perimeter is 264, find the area.	9. The width of a rectangle is 5 more than its length. Find the perimeter if the area is 150.
4. The width of a rectangle is 7 more than twice the length. Find the area if the perimeter is 146.	10. The length of a rectangle is triple its width. Find the perimeter if the area is 243.
5. The length of a rectangle is five times its width. If the perimeter is 360, find the area.	11. The length of a rectangle is one more than twice its width. Find the perimeter if the area is 253.
6. The length of a rectangle is triple its width. If the perimeter is 264, find the area.	12. The length of a rectangle is two less than triple its width. Find the perimeter if the area is 341.

Rhombus 1

Find the area of the rhombus in square units.	Find the height of the rhombus.
1. diagonals 14 and 35	8. diagonals 18 and 24
2. perimeter 80, one angle 60°	9. perimeter 44, area 165
3. perimeter 52, shorter diagonal 10	10. diagonals 20 and 48
4. diagonals 22 and 47	11. shorter diagonal 10, one angle 120°
5. perimeter 60, longer diagonal 24	12. perimeter 100, area 375
6. perimeter 20, longer diagonal 8	13. diagonals 14 and 48
7. diagonals 22 and 38	14. one angle 45°, perimeter $4\sqrt{2}$

Rhombus 2

Find the perimeter of the rhombus.	Find the diagonal(s) of the rhombus.
1. diagonals 24 and 32	7. perimeter 60, longer diagonal 24
2. longer diagonal 18, one angle 60°	8. perimeter 32, one diagonal 10
3. area 1320, shorter diagonal 22	9. area 720, perimeter 164
4. diagonals 30 and 72	10. one diagonal $8\sqrt{2}$, perimeter $16\sqrt{5}$
5. diagonals 10 and 14	11. area 840, perimeter 116
6. one angle 135°, area $72\sqrt{2}$	12. perimeter 20, ratio of diagonals 2:1

Scientific Notation 1

Write in scientific notation.

1. 5468	17. 54,931	33. .4545 x 10^{-4}
2. 29	18. 44 x 10^3	34. .0078
3. 0.932	19. .000071	35. 5^2 x 3^2
4. 80,022	20. .4378	36. 12 x 400
5. 321	21. 12,968	37. 58,932 ÷ 100
6. 56 x 10	22. 22.675 x 10^4	38. 10,001
7. 895.2 x 10^6	23. 76.45 ÷ 100	39. 157.9 x 10^7
8. 553 x 10^{-3}	24. .0045	40. 36 x 11
9. 321 ÷ 1000	25. 36 x 10^{-4}	41. .000422
10. .72 x 10^3	26. 623 x 10^7	42. 85 x 10^2
11. 50 x 15	27. 387.93	43. 675.8 x 10^5
12. 0.99	28. 1911.7	44. 903.6
13. 10,854	29. 12 x 12	45. 20 x 30 x 40
14. 6519 ÷ 100	30. 78,935	46. .97 x 10^{-3}
15. 42.35 x 10^4	31. 25 x 10^5	47. .0265
16. 300 x 14	32. 5370 ÷ 10	48. 4^2 x 2^5

Scientific Notation 2

Write in scientific notation.

1. $(7 \times 10^{-6}) \div (8 \times 10^{-4})$	11. $(2.5 \times 10^{10}) \times (5 \times 10^{-3})$
2. $(5 \times 10^{-8}) \times (9 \times 10^{2})$	12. $(451 \times 10^{6}) \div (11 \times 10^{3})$
3. $(42 \times 10^{15}) \div (4 \times 10^{13})$	13. $(25 \times 10^{11}) \div (2 \times 10^{8})$
4. $(2 \times 10^{9}) \div (8 \times 10^{3})$	14. $(3 \times 10^{9}) \div (4 \times 10^{-3})$
5. $(9 \times 10^{-3}) \times (9 \times 10^{11})$	15. $(5 \times 10^{-2}) \times (7 \times 10^{13})$
6. $(64 \times 10^{3}) \div (5 \times 10^{6})$	16. $(253 \times 10^{-7}) \div (1.1 \times 10^{5})$
7. $(3 \times 10^{5}) \div (8 \times 10^{-5})$	17. $(1 \times 10^{5}) \div (8 \times 10^{-1})$
8. $(6 \times 10^{-3}) \times (6 \times 10^{-7})$	18. $(11 \times 10^{2}) \times (36 \times 10^{-17})$
9. $(5 \times 10^{4}) \div (8 \times 10^{-3})$	19. $(2.5 \times 10^{-2}) \div (10 \times 10^{5})$
10. $(3 \times 10^{2}) \div (12 \times 10^{-9})$	20. $(4.5 \times 10^{9}) \div (9 \times 10^{-1})$

Sequences 1

Find the next term, write the key, and circle A for arithmetic, G for geometric, or N for neither.

1. 6, 9, 12, 15, 18, _____

    A   G   N   Key: _____

2. 3, 12, 48, _____

    A   G   N   Key: _____

3. 5, −1, −7, −13, −19, _____

    A   G   N   Key: _____

4. 1, 4, 9, 16, 25, _____

    A   G   N   Key: _____

5. 2, 3, 5, 7, 11, 13, 17, _____

    A   G   N   Key: _____

6. 3, 10, 31, 94, _____

    A   G   N   Key: _____

7. 1, 3, 6, 10, 15, 21, 28, _____

    A   G   N   Key: _____

8. 300, 150, 75, 37.5, _____

    A   G   N   Key: _____

9. 6, −12, 24, −48, 96, _____

    A   G   N   Key: _____

10. 1, 1, 2, 3, 5, 8, 13, 21, 34, _____

    A   G   N   Key: _____

11. 3, 14, 58, 234, _____

    A   G   N   Key: _____

12. 1, 8, 27, 64, 125, _____

    A   G   N   Key: _____

13. 1, 4, 16, 64, _____

    A   G   N   Key: _____

14. 3, 7, 15, 31, 63, _____

    A   G   N   Key: _____

15. 5, 15, 45, 135, _____

    A   G   N   Key: _____

16. 1, 2, 6, 24, 120, _____

    A   G   N   Key: _____

17. 500, 100, 20, 4, 0.8, _____

    A   G   N   Key: _____

18. 3.3, 4.5, 5.7, 6.9, _____

    A   G   N   Key: _____

19. 13, 16, 22, 31, 43, _____

    A   G   N   Key: _____

20. 13, 4, −5, −14, −23, _____

    A   G   N   Key: _____

21. 6, 11, 21, 41, _____

    A   G   N   Key: _____

22. 20, 30, 50, 70, 110, 130, _____

    A   G   N   Key: _____

Sequences 2

Find the specified ordinal term.

1. 1900th of ABCDEABCDE . . .	12. 1466th of 76543217654321 . . .
2. 942nd of ABCDABCD . . .	13. 767th of +&*@!%$+&*@!%$ . . .
3. 1663rd of DEFGHDEFGH . . .	14. 598th of BOPHBOPH . . .
4. 389th of @#$%&@#$%& . . .	15. 5392nd of SWAESWAE . . .
5. 355th of ?+¢?+¢ . . .	16. 1937th of B9X5B9X5 . . .
6. 667th of ABWXYZABWXYZ . . .	17. 6431st of KLHGTKLHGT . . .
7. 557th of N8M9PN8M9P . . .	18. 742nd of ≠∆^≠∆^≠∆^ . . .
8. 843rd of 1234567812345678 . . .	19. 797th of ABABAB . . .
9. 758th of abcdefgabcdefg . . .	20. 8613rd of @%$@%$@%$ . . .
10. 9572nd of ßπ∂ßπ∂ßπ∂ . . .	21. 2146th of mnopqswmnopqsw . . .
11. 6583rd of ©®∑§©®∑§ . . .	22. 7898th of SEDRTSEDRT . . .

Sequences 3

Find the ones digit.

1. 97^{241}	12. 264^{265}	23. 293^{255}
2. 2^{391}	13. 624^{326}	24. 143^{512}
3. 293^{615}	14. 358^{170}	25. 17^{902}
4. 84^{781}	15. 172^{379}	26. 118^{987}
5. 37^{362}	16. 437^{272}	27. 355^{255}
6. 325^{506}	17. 93^{762}	28. 552^{988}
7. 189^{137}	18. 397^{175}	29. 559^{154}
8. 78^{765}	19. 532^{197}	30. 128^{521}
9. 237^{236}	20. 106^{234}	31. 337^{333}
10. 112^{137}	21. 268^{268}	32. 353^{422}
11. 117^{183}	22. 802^{650}	33. 651^{156}

Sequences 4

Find the specified term of the arithmetic sequence.

1. 4, 7, 10, 13, . . . Find 20th term.	12. 40, 50, 60, 70, . . . Find 111th term.
2. 4, 4.5, 5, 5.5, . . . Find 20th term.	13. 2, 7, 12, 17, . . . Find 45th term.
3. 8, 10, 12, 14, . . . Find 57th term.	14. 15, 30, 45, 60, . . . Find 41st term.
4. 50, 45, 40, 35, . . . Find 19th term.	15. 17, 11, 5, −1, . . . Find 81st term.
5. 80, 40, 0, −40, . . . Find 11th term.	16. −8, −5, −2, 1, . . . Find 17th term.
6. 33, 44, 55, 66, . . . Find 26th term.	17. 9, 20, 31, 42, . . . Find 28th term.
7. 40, 36, 32, 28, . . . Find 43rd term.	18. 60, 25, −10, −45, . . . Find 11th term.
8. 11, 17, 23, 29, . . . Find 51st term.	19. 12, 37, 62, 87, . . . Find 10th term.
9. 13, 21, 29, 37, . . . Find 26th term.	20. 40, 29, 18, 7, . . . Find 16th term.
10. 28, 25, 22, 19, . . . Find 32nd term.	21. −20, −10, 0, 10, . . . Find 14th term.
11. 30, 38, 46, 54, . . . Find 71st term.	22. −7, −3, 1, 5, . . . Find 21st term.

Sequences 5

Find the next term in each sequence by successive differences.

1.	7	7	11	20	37	69	129	238

2.	0	12	25	41	62	90	127

3.	5	6	14	32	64	115	191

4.	6	11	20	35	60	101	166	265

5.	11	21	39	69	116	188	298	466

6.	11	21	40	69	112	176	271	410

7.	18	28	40	56	78	108	148	200

Sequences 6

Answer as indicated.	Find the specified digit.
1. 90, A, 76, 69, 62, B, 48, C, 34 Find A + B + C.	12. 18th decimal digit of 3/11
2. 3, A, 19, B, 35, 43, 51 Find AB.	13. 37th decimal digit of 1/7
3. 3, 15, 75, A, 1875, B Find B − A.	14. 64th decimal digit of 2/7
4. 5, A, 17, 23, B, C, 41, 47 Find A + B + C.	15. 85th decimal digit of 1/18
5. 8, 20, 50, A, B, 781.25 Find A + B.	16. 44th decimal digit of 3/7
6. 8, 19, A, 41, 52, B, C, 85 Find C(B − A).	17. 98th decimal digit of 1/37
7. A, 1600, 1200, 800, B Find B − A.	18. 54th decimal digit of 2/37
8. 60, 55.5, 51, A, B, 37.5, 33, C Find C − (A + B).	19. 76th decimal digit of 4/37
9. 2, 6, 18, 54, A, B, 1458 Find B − A.	20. 35th decimal digit of 5/7
10. 1000, 200, 40, A, 1.6, B Find A + B.	21. 64th decimal digit of 6/7
11. 6, 9.5, A, B, 20, 23.5, 27, C Find A + B + C.	22. 100th decimal digit of 1/36

Sequences 7

List the first 5 elements of the linear sequence for whole number input.	Find the formula in functional notation with whole number input for the sequence.
1. $f(n) = 3n - 4$	14. $-2, 3, 8, 13, 18, \ldots$
2. $f(n) = -5n + 6$	15. $-7, -3, 1, 5, 9, \ldots$
3. $f(n) = 2n - 7$	16. $-1, 5, 11, 17, 23, \ldots$
4. $f(n) = -4n - 2$	17. $-3, -1, 1, 3, 5, \ldots$
5. $f(n) = 6n + 1$	18. $-8, -5, -2, 1, 4, \ldots$
6. $f(n) = 7n + 4$	19. $-4, 3, 10, 17, 24, \ldots$
7. $f(n) = 8n - 5$	20. $6, 10, 14, 18, 22, \ldots$
8. $f(n) = 5n - 6$	21. $12, 6, 0, -6, -12, \ldots$
9. $f(n) = 9n + 3$	22. $8, 11, 14, 17, 20, \ldots$
10. $f(n) = -3n + 7$	23. $10, 6, 2, -2, -6, \ldots$
11. $f(n) = -9n - 8$	24. $5, -2, -9, -16, -23, \ldots$
12. $f(n) = -8n + 9$	25. $-5, -9, -13, -17, -21, \ldots$
13. $f(n) = -2n - 1$	26. $-6, 3, 12, 21, 30, \ldots$

Sequences 8

Find the first term a_1 of the arithmetic sequence given two terms.	*Find the first term a_1 and the common difference d of the arithmetic sequence.*
1. $a_6 = 24$ $a_{20} = 80$	11. sum of 1st 20 terms = 1280 $a_{20} = 121$
2. $a_5 = -13$ $a_{30} = 62$	12. sum of 1st 25 terms = 950 $a_{25} = 92$
3. $a_8 = -10$ $a_{41} = 221$	13. sum of 1st 37 terms = 5439 $a_{37} = 282$
4. $a_7 = 9$ $a_{35} = 233$	14. sum of 1st 31 terms = 1798 $a_{31} = 166$
5. $a_{11} = 44$ $a_{45} = -58$	15. sum of 1st 19 terms = 2356 $a_{19} = 232$
6. $a_9 = 31$ $a_{53} = 97$	16. sum of 1st 17 terms = 1700 $a_{17} = 156$
7. $a_{12} = -12$ $a_{42} = 60$	17. sum of 1st 21 terms = 1281 $a_{21} = 111$
8. $a_{13} = 10$ $a_{33} = 80$	18. sum of 1st 15 terms = 1245 $a_{15} = 146$
9. $a_{10} = 195$ $a_{55} = 420$	19. sum of 1st 45 terms = 3195 $a_{45} = 137$
10. $a_{20} = 150$ $a_{58} = 378$	20. sum of 1st 13 terms = 806 $a_{13} = 110$

Series 1

Find the sum by the formula (F + L)n/2.

1. $20 + 21 + 22 + \ldots + 78 + 79 + 80$	9. $11 + 13 + 15 + \ldots + 87 + 89 + 91$
2. $16 + 20 + 24 + \ldots + 64 + 68 + 72$	10. $15 + 20 + 25 + \ldots + 75 + 80 + 85$
3. $18 + 24 + 30 + \ldots + 84 + 90 + 96$	11. $31 + 38 + 45 + 52 + 59 + 66 + 73$
4. $100 + 105 + 110 + \ldots + 245 + 250$	12. $300 + 301 + \ldots + 499 + 500$
5. $32 + 33 + 34 + \ldots + 76 + 77 + 78$	13. $44 + 50 + 56 + \ldots + 86 + 92 + 98$
6. $28 + 29 + 30 + \ldots + 84 + 85 + 86$	14. $6 + 9 + 12 + \ldots + 66 + 69 + 72$
7. $2 + 4 + 6 + \ldots + 64 + 66 + 68$	15. $1 + 8 + 15 + \ldots + 78 + 85 + 92$
8. $50 + 52 + 54 + \ldots + 94 + 96 + 98$	16. $1 + 2 + 3 + \ldots + 198 + 199 + 200$

Series 2

Find the sum by the formula for consecutive odds starting at one.

1. the first 7 odd numbers	11. 1 + 3 + 5 + 7 + . . . + 49
2. the first 11 odd numbers	12. 1 + 3 + 5 + 7 + . . . + 79
3. the first 15 odd numbers	13. 1 + 3 + 5 + 7 + . . . + 119
4. the first 20 odd numbers	14. 1 + 3 + 5 + 7 + . . . + 149
5. the first 49 odd numbers	15. 1 + 3 + 5 + 7 + . . . + 179
6. the first 70 odd numbers	16. 1 + 3 + 5 + 7 + . . . + 199
7. the first 80 odd numbers	17. 1 + 3 + 5 + 7 + . . . + 249
8. the first 91 odd numbers	18. 1 + 3 + 5 + 7 + . . . + 299
9. the first 100 odd numbers	19. 1 + 3 + 5 + 7 + . . . + 399
10. the first 200 odd numbers	20. 1 + 3 + 5 + 7 + . . . + 499

Sets 1

Check for yes and dash for no if the number belongs to the set.

1.

	25	1.5	$\frac{2}{7}$	$|-2|$	0	$\sqrt{16}$	$\frac{-6}{3}$	$\sqrt{-36}$	-17	$9.\overline{4}$	$9\frac{3}{11}$
N											
W											
Z											
Q											
R											

2.

	$\frac{3}{5}$	π	$\sqrt{81}$	64	$2.\overline{14}$	-0.6	$-|-8|$	-1	$27\frac{4}{5}$	$\frac{56}{7}$	$\sqrt{-4}$
N											
W											
Z											
Q											
R											

3.

	-8	$12\frac{3}{8}$	$\frac{63}{9}$	$37.\overline{11}$	$\sqrt{-9}$	121	$\frac{1}{4}$	$\sqrt{49}$	.053	4π	$|2-3|$
N											
W											
Z											
Q											
R											

4.

	12.8	$|-45|$	$\sqrt{-26}$	-81	$\sqrt{144}$	3.888	$\frac{35}{7}$	$10.\overline{64}$	100	$\frac{4}{9}$	$33\frac{1}{3}$
N											
W											
Z											
Q											
R											

Sets 2

Complete with the set name and/or operation(s). Also name the property.	Find the intersection.
1. A∪B = B∪ ___	17. {primes} ∩ {multiples of 11}
2. B∩C = ___ ∩B	18. {zero} ∩ {evens}
3. (A∪F)∪H = A∪(F ___ H)	19. {odds} ∩ {evens}
4. M∪(___ ∩ ___) = (M∪N)∩(M∪T)	20. {primes} ∩ {factors of 30}
5. F ___ (C∩D) = (F∩C) ___ D	21. {factors of 36} ∩ {multiples of 4}
6. G∪B = B ___ G	22. {2-digit product of 2 primes} ∩ {factors of 330}
7. B∩(C∩D) = (B∩C)∩ ___	23. {composites} ∩ {factors of 20}
8. Y∩ ___ = X∩Y	24. {perfect squares} ∩ {factors of 500}
9. ___ ∪S = S∪T	25. {perfect squares} ∩ {2-digit perfect cubes}
10. (D ___ F)∪S = D∪(F ___ S)	26. {primes} ∩ {evens}
11. K∩(L∪M) = (K ___ L)∪(K ___ M)	27. {x-axis ordered pairs} ∩ {y-axis ordered pairs}
12. (___ ∩H)∩K = J∩(H∩K)	28. {x|x ε N, 2x < 8 } ∩ {x|x ε N, 3 ≤ 3x ≤ 9}
13. (B∪C)∩D = (B∩D)∪(C∩ ___)	29. {x|x ε Z, 4x < 28 } ∩ {x|x ε Z, 3x > 9}
14. ___ ∪(C∪E) = (B∪C)∪E	30. {x|x ε W, 3x < 9 } ∩ {x|x ε W, −2 < 7x < 20}
15. X∪(S ___ T) = (X∪S) ___ (X∪T)	31. {x|x ε Z, 5x ≥ 25 } ∩ {x|x ε Z, −5 ≤ 3x < 25}
16. F ___ (G∪H) = (F ___ G)∪(___ ∩H)	32. {x|x ε N, 4x ≤ 9 } ∩ {x|x ε N, 3x + 4 > 9}

Sets 3

Operate as indicated.

A = {1, 2, 5, 7} C = {1, 3, 5, 7} B = {2, 4, 7, 8} D = {2, 4, 6, 8} U = {1, 2, 3, 4, 5, 6, 7, 8, 9}	A = {1, 2, 3, 5, 6} C = {1, 3, 5, 7} B = {2, 4, 5, 8} D = {1, 3, 9} U = {1, 2, 3, 4, 5, 6, 7, 8, 9, 10}
1. A ∪ C	10. (A ∪ B) ∪ C
2. C ∩ D	11. (A ∩ B) ∩ C
3. A ∩ C	12. A ∪ (B ∩ C)
4. C′	13. A ∩ (B ∪ D)
5. B′ ∪ C	14. (A ∪ D) ∪ B′
6. B ∪ (D′ ∩ A)	15. (A ∪ C) ∩ D′
7. (A ∪ B) ∩ (C ∪ D)	16. (C′ ∪ B) ∪ A
8. A ∩ (D′ ∪ C)	17. (B ∪ D) ∪ A′
9. (B′ ∩ C) ∪ (A′ ∩ D)	18. D ∩ (A ∪ C′)

Sets 4

Draw a Venn diagram of the sets to show their relationship.

1.　A = {2, 4, 6, 8}　B = {2, 3, 4, 5}	7.　A = {1, 2, 3, 4} 　　B = {5, 6} 　　C = {1, 2, 5, 6, 7, 8}
2.　A = {1, 2, 3, 4}　B = {5, 6, 7}	8.　A = {2, 8, 9}　B = {1, 4, 5, 7} 　　C = {0, 4, 5, 6}
3.　A = {1, 2, 3, 4, 5}　B = {2, 4, 5}	9.　A = {0, 1, 5}　B = {3, 7, 8} 　　C = {0, 1, 2, 4, 5, 6}
4.　A = {1, 2, 3, 4, 5}　B = {3, 5, 7} 　　C = {6, 7, 8, 9}	10.　A = {1, 2, 3, 6, 7} 　　B = {1, 2} 　　C = {1, 2, 3, 4, 5}
5.　A = {0, 3, 4, 7}　B = {1, 3, 4} 　　C = {2, 5}	11.　A = {1, 2, 9}　B = {3, 4, 5} 　　C = {0, 6, 7, 8}
6.　A = {2, 4, 6, 8} 　　B = {3, 4, 8, 9} 　　C = {1, 6, 7, 8, 9}	12.　A = {1, 3, 4, 6, 9} 　　B = {0, 1, 2, 8} 　　C = {1, 2, 6, 9}

Sets 5

Operate on the set of points in the figures.

1. $\overline{AB} \cap \overline{BC}$	14. $\overline{AD} \cap \overline{GH}$	27. $\overleftrightarrow{AB} \cap \overline{BC}$
2. $\overline{AC} \cap \overline{BD}$	15. $\overline{AD} \cup \overline{GH}$	28. $\overline{AB} \cap \overline{DC}$
3. $\overline{AE} \cup \overline{EC}$	16. $\overline{EH} \cap \overline{HC}$	29. $\overrightarrow{BA} \cap \overrightarrow{BC}$
4. $\overrightarrow{DA} \cup \overrightarrow{DC}$	17. $\overline{AE} \cup \overline{EH} \cup \overline{AH}$	30. $\overline{AB} \cup \overline{BC}$
5. $\overline{BE} \cap \overline{AD}$	18. $\overrightarrow{GH} \cup \overrightarrow{AH}$	31. $\overline{BC} \cup \overrightarrow{CD}$
6. $\overline{AB} \cup \overline{BE} \cup \overline{AE}$	19. $\overline{AG} \cup \overline{GH} \cup \overline{HD}$	32. $\overline{AD} \cap \overline{AB}$
7. $\overline{BC} \cap \overline{CD}$	20. $\overline{AB} \cap \overline{CD}$	33. $\overrightarrow{AC} \cap \overrightarrow{DC}$
8. $\overline{AB} \cap \overline{CD}$	21. $\overrightarrow{CF} \cup \overrightarrow{CE}$	34. $\overline{AC} \cup \overline{CD}$
9. $\overrightarrow{EB} \cup \overrightarrow{EC}$	22. $\overline{AG} \cup \overline{GH}$	35. $\overrightarrow{AB} \cap \overline{BD}$
10. $\overrightarrow{DB} \cap \overrightarrow{EB}$	23. $\overline{AE} \cap \overline{AF}$	36. $\overline{CD} \cap \overline{AD}$
11. $\overrightarrow{EC} \cup \overrightarrow{EA}$	24. $\overrightarrow{FG} \cap \overrightarrow{GC}$	37. $\overrightarrow{BA} \cup \overrightarrow{BC}$
12. $\overline{BD} \cap \overline{CD}$	25. $\overline{GH} \cup \overline{HC} \cup \overline{GC}$	38. $\overrightarrow{AB} \cup \overrightarrow{CD}$
13. $\overline{BC} \cup \overline{BD} \cup \overline{CD}$	26. $\overrightarrow{GA} \cap \overrightarrow{HD}$	39. $\overrightarrow{AD} \cap \overline{BC}$

Sets 6

Operate on the infinite sets.	Find the number of subsets.
1. Z ∪ W = _____	17. {Bob, Joe}
2. N ∩ W = _____	18. {0, 1, 2, 3, 4, 5, 6, 7, 8}
3. W ∩ Q = _____	19. {H, E, L, L, O}
4. N ∪ Z = _____	20. {10, 20, 30}
5. Q ∩ Z = _____	21. {n, u, m, b, e, r}
6. R ∩ Q = _____	22. {c, l, a, p} ∪ {h, a, n, d, s}
7. N ∪ Q = _____	23. {c, o, u, n, t}
8. Z ∪ (Q ∪ N) = _____	24. {s, e, t} ∪ {t, h, e, o, r, y}
9. R ∩ (N ∪ W) = _____	25. {M, A, T, H}
10. Q ∩ (N ∪ Z) = _____	26. {A, L, G, E, B, R, A}
11. (Q ∪ Z) ∩ R = _____	27. {h, o, u, s, e} ∪ {b, o, a, t}
12. (Z ∪ R) ∪ W = _____	28. {1, 3, 5, 7} ∪ {2, 4, 6, 8}
13. Z ∩ (Q ∪ R) = _____	29. {9, 8, 7, 6, 5, 4, 3, 2, 1, 0}
14. W ∩ (Q ∪ N) = _____	30. {3, 4, 5, 6} ∪ {6, 7, 8, 9}
15. Z ∪ (W ∩ R) = _____	31. {v, a, l, u, e}
16. W ∪ (Q ∩ N) = _____	32. {23, 24, 24, . . . , 33}

Spheres 1

Find the volume and surface area of the sphere with the specified dimension.

1. r = 3	9. d = 8
2. d = 18	10. r = 3.5
3. r = 1.5	11. r = 5
4. d = 5	12. r = 1/3
5. r = 6	13. d = 30
6. r = 1	14. r = .25
7. r = 2	15. d = 24
8. d = 1	16. d = 9

Spheres 2

Find the ratio.

1. Find the ratio of the volumes of two spheres whose surface areas are 36 and 64.

2. Find the ratio of the volumes of two spheres whose surface areas are 12 and 27.

3. Find the ratio of the surface areas of two spheres with volumes 64 and 729.

4. Find the ratio of the surface areas of 2 spheres whose volumes are in the ratio 512:1331.

5. Find the ratio of the surface areas of 2 spheres whose volumes are in the ratio 343:1728.

6. Find the ratio of the volumes of two spheres whose surface areas are 20 and 245.

7. Find the ratio of the surface areas of 2 spheres whose volumes are in the ratio 24:81.

8. Find the ratio of the volumes of two spheres whose surface areas are in the ratio 12:75.

9. Find the ratio of the surface areas of 2 spheres whose volumes are in the ratio 8:243.

10. Find the ratio of the volumes of two spheres whose surface areas are in the ratio 81:121.

Spheres 3

Answer as indicated about inscribed spheres.

Answer as indicated, converting volume and surface area of spheres.

1. Find the volume of a sphere inscribed in a cube with edge 6.

7. A sphere has surface area 36π. Find its volume.

2. Find the volume of a sphere inscribed in a cube with edge 11.

8. A sphere has surface area 324π. Find its volume.

3. A sphere just fits in a cylinder with volume 54π. Find the surface area of the cylinder.

9. A sphere has volume 288π. Find its surface area.

4. A cylinder circumscribed about a sphere has volume 128π. Find the volume of the sphere.

10. A sphere has volume $36,000\pi$. Find its surface area.

5. Three spheres each with radius 10 just fit inside a cylinder. Find the volume and surface area of the cylinder.

11. A sphere has surface area 576π. Find its volume.

6. A sphere with radius 6 just fits in a cylinder. Find the surface area of the cylinder.

12. A sphere has volume 4500π. Find its surface area.

Spheres 4

Find the total volume or surface area of the shape, attaching base to base.	Answer about reshaped spheres.
1. A hemisphere is attached to one base of a cylinder with radius 3 and height 8. Find the volume.	6. A meatball with a 3-inch radius can be reformed into how many meatballs with a 1-inch radius?
2. Use the same shape as #1. Find the surface area.	7. Eight metal balls each with radius 1 are melted and reshaped into one ball. Find the new radius.
3. A hemisphere is attached to the base of a cone with height 11 and radius 5. Find the volume.	8. How many spherical snowballs with a 1-inch radius can be made from a 6-inch radius spherical snowball?
4. A hemisphere attached to each base of a cylinder. The completed shape has diameter 12 and total length 26. Find the volume.	9. How many spherical snowballs with a 3-inch radius can be made from a 1-foot radius spherical snowball?
5. Use the same shape as #4. Find the surface area.	10. How many spherical snowballs with a 3-inch radius can be made from a 2-foot radius spherical snowball?

Statistics 1

Answer as indicated.

1. Of 5 test scores from 0 to 100, the mode is 80, the median is 75, the mean is 69, and the range is 35. Find the next to the least.	6. Of a set of seven 2-digit positive multiples of 10, 30 is the median, 20 and 40 are the modes, and 110 is the range. Find the mean.
2. In a set of 5 numbers, the mode is less than the median. If four of the numbers are 15, 35, 45, and 65, find the missing number.	7. Of 5 numbers from 10 to 50, the mean is 33, the median is 30, the mode is 48, and the range is 37. Find the second least.
3. In a set of 5 numbers, the mean, median, and mode are all equal. If 4 of the numbers are 50, 60, 40, and 10, find the missing number.	8. Of 6 test scores from 0 to 100, the modes are 60 and 80, the median is 65, and the range is 70. Find the mean.
4. Five natural numbers have mean 5, median 5, and mode 8. Find the range.	9. The mode of x, $4x$, -12, 18, and $10x$ is 18. If the median is 7.2, find the mean.
5. Of a set of seven 2-digit multiples of 5, the range is 85, the median is 25, the modes are 10 and 25, and the mean is 40. Find the sum of the 3 greatest.	10. In a set of 5 numbers, the mean, median, and mode are all equal. If 4 of the numbers are 18, 29, 21, and 16, find the missing number.

Statistics 2

Find the mean of:	Answer as indicated.
1. 5x + 1, 3x − 2, 6x + 7, and 2x + 10.	8. Find the mean, median, and range of the factors of 24.
2. −3x + 5, −x − 7, 11x + 6, 2x − 12, 4x, and 8x + 17.	9. Find the mean, median, mode, and range of 10 numbers: the 1st 5 squares and the 1st 5 cubes.
3. 6x + 4, 4x − 3, 7x + 8, x − 11, and 2x + 17.	10. Find the mean, median, and range of the factors of 36.
4. −8x + 6, −4x − 5, 10x + 9, 3x − 17, 8x + 1, 3x, and 9x + 6.	11. Find the mean, median, and range of the arithmetic sequence with F = 2, L = 20, and d = 3.
5. −2x + 8, −7x − 9, 12x + 4, 4x − 10, 5x + 5, 9x, x + 1, and −8x + 13.	12. Find the mean, median, and range of the arithmetic sequence with F = 2, L = 200, and d = 2.
6. −4x + 6, −2x − 9, 13x + 4, 2x − 11, 6x, 5x + 1, 2x − 6, −x − 8, and 9x + 14.	13. Find the mean, median, and range of the first 8 prime numbers.
7. −x + 8, −5x − 9, 14x + 5, 3x − 10, 8x + 3, and 7x + 7.	14. Find the mean, median, and range of the 3-digit multiples of 50.

Statistics 3

Find the positive difference of the median and mode.

1.
stem	leaf
2	1 1 3 4 7 7
3	0 6 6 6 8
5	3 3 6 6 8 9
6	1 2 4 4
8	0 1 3
9	8

6.
```
X
X           X           X
X      X  X             X
X  X  X  X              X
X  X  X  X  X  X
X  X  X  X  X  X
3  4  5  6  7  8
```

2.
stem	leaf
1	0 0 0 6 7 9 9
2	1 1 2 3 4 5 6
4	2 2 3 7 7
6	3 3 3 4 5 6
7	0 1
8	3 3 5 5 5 5

7.
```
X
X               X          X  X
X         X  X             X  X
X  X  X  X                 X  X
X  X  X  X  X     X  X  X
X  X  X  X  X     X  X  X
0  2  4  6  8  10 12
```

3.
stem	leaf
1	1 2 2 4 4 6 8
2	0 0
4	0 1 2 3 3 5
6	2 8 9
7	3 3 3 5 7
8	0 0 0
9	1 2 2 2 2 5 7 8 9

8.
score	frequency
100	5
90	10
80	15
70	20
60	5
50	15
40	15

4.
stem	leaf
1	0 0 1 1 2
2	2 3 3
3	3 4 4 4
4	5 6 6 7 8
5	5 5
6	1 1 2 2 5 5 5 5 6
7	0 3 6

9.
score	frequency
70	12
60	22
50	18
40	20
30	12
20	10
10	8

5.
stem	leaf
3	1 1
4	3 4 5 6
5	0 0 9 9 9 9 9
6	0 2 2 4 4 5 5 6 6
7	0 0 0 0 0 2 3 4 5 6
8	1 1 1 1 1 1 2 2 2 4 5
9	5 5

10.
score	frequency
100	20
110	30
120	11
130	16
140	19
150	40
160	10

Statistics 4

Answer as indicated.

1. The median of 11, 42, 65, *x*, and 16 is 8 less than the mean. Find *x* if *x* is negative.	6. The mean of a set of 3 test scores is 90, the median is 88, and the greatest is 97. Find the least score.
2. The median of 18, 32, *x*, 12, 50, and 15 is 15 less than the mean. Find *x* if *x* is 3-digit and positive.	7. The mean of a set of 3 test scores is 85, the median is 82, and the least is 77. Find the greatest score.
3. The median of 30, 12, *x*, 15, 40, and 17 is 10 more than the mean. Find *x* if *x* is negative.	8. Of 4 test scores, the mean is 81, the median is 80, the mode is 73, and the least is 73. Find the greatest score.
4. The median of 9, 2, *x*, 4, 3, 9, 7, and 5 is 9 less than the mean. Find *x* if *x* is 2-digit and positive.	9. Of 4 test scores, the mean is 78, the median is 79, the mode is 86, and the greatest is 86. Find the least score.
5. The median and mode of 8, *x*, 2.5*x*, 3*x*, and 10 are both 10. Find the sum of possible means if all five numbers are natural.	10. The median and mode of 7, *x*, 1.5*x*, 2*x*, and 8 are both 8. Find the sum of possible means if all five numbers are natural.

Transformations 1

Draw the reflection over the specified line. One box equals one unit.

1. over the *y*-axis

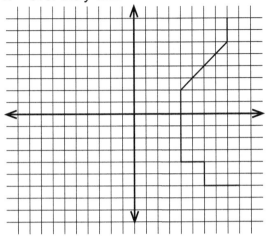

4. over the line *x* = 2

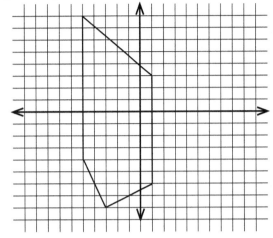

2. over the *x*-axis

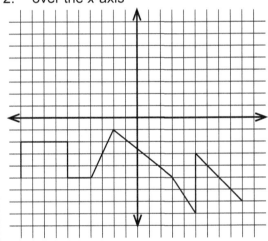

5. over the line x = −1

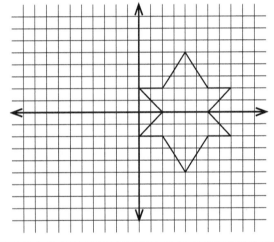

3. over the line *y* = −1

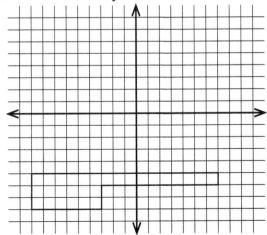

6. over the *x*-axis

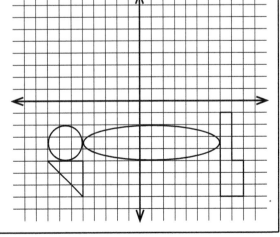

Transformations 2

Find the coordinates of the point after the successive reflection and translation.	Find the coordinates of the point after the successive translation and reflection.
1. (1, 4) across the x-axis 2 units left	17. (3, 1) 4 units right across the y-axis
2. (2, 7) across the y-axis 9 units down	18. (−3, 5) 6 units up across the x-axis
3. (−1, 5) across y = 2 5 units right	19. (−4, 6) 7 units left across x = 4
4. (−7, −2) across y = x 4 units up	20. (8, −9) 3 units up across y = −x
5. (4, −5) across the y-axis 8 units down	21. (−5, 6) 10 units down across the x-axis
6. (1, 3) across x = 8 14 units left	22. (−1, −8) 12 units right across y = −3
7. (8, 1) across y = −x 4 units up	23. (0, 4) 11 units down across y = x
8. (10, 4) across y = −1 7 units right	24. (7, 3) 9 units left across x = −6
9. (9, −2) across the x-axis 5 units down	25. (−6, 5) 8 units up across the y-axis
10. (−3, 8) across x = 1 8 units left	26. (−4, 11) 2 units right across y = 7
11. (7, 3) across the y-axis 2 units right	27. (2, −3) 5 units left across the x-axis
12. (−5, 6) across y = 3 12 units up	28. (−3, 12) 15 units down across x = 5
13. (12, 10) across y = x 6 units down	29. (9, 8) 14 units up across y = −x
14. (−2, 9) across the x-axis 9 units right	30. (−7, 8) 6 units left across the y-axis
15. (−4, 11) across y = 4 7 units up	31. (−7, 0) 4 units down across x = 2
16. (0, −6) across y = −x 10 units left	32. (−5, −1) 13 units right across y = x

Transformations 3

Draw the reflection of each shape over the origin.

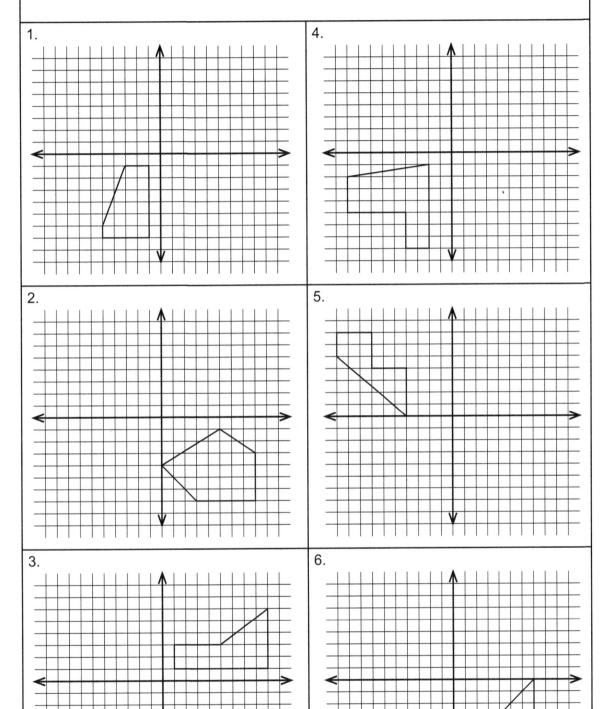

1.

2.

3.

4.

5.

6.

Transformations 4

Name the fewest transformation(s)–rotation, reflection, and/or translation–done to the left figure to obtain the right figure.

1.

2.

3.

4.

5.

6.

7.

8.

Trapezoids 1

Find the midline or base of the trapezoid by mental math.

	b	M	B		b	M	B		b	M	B
1.	10		20	17.	6.25	18.5		33.		40	47
2.	8		24	18.	7	15		34.		22	26.1
3.	8.2		10.3	19.	9	29.5		35.		35.6	38.6
4.	11		13	20.	64	64.5		36.		84	95
5.	29		41	21.	35	40		37.		11	15.8
6.	10		15	22.	2.3	4		38.		13.8	14
7.	9.5		15.5	23.	12.4	24		39.		55	59
8.	36		52	24.	12.5	15.3		40.		70	83
9.	19		42	25.	22.8	23.5		41.		63	90
10.	11		20	26.	18	24.2		42.		18.4	21.8
11.	16.5		23.5	27.	53	58.5		43.		29	43
12.	13		25	28.	11.1	17.2		44.		51	59
13.	15		25	29.	30.4	35.4		45.		14.5	27.5
14.	12.5		18.7	30.	55	72.5		46.		33	37.2
15.	4.1		7.9	31.	16.3	20.4		47.		16.2	17.3
16.	22.6		32.4	32.	5.7	5.75		48.		20	33

Trapezoids 2

Find the area of the trapezoid in square units by mental math given the bases and height.

1.	b = 12, B = 18 H = 9	17.	b = 59, B = 65 H = 11
2.	b = 6.5, B = 9.5 H = 7	18.	b = 22, B = 66 H = 5
3.	b = 7, B = 15 H = 11	19.	b = 17, B = 30 H = 20
4.	b = 6.25, B = 11.75 H = 7	20.	b = 11.2, B = 18.8 H = 15
5.	b = 11, B = 14 H = 10	21.	b = 18, B = 34 H = 25
6.	b = 13, B = 19 H = 15	22.	b = 13, B = 19 H = 15
7.	b = 10.1, B = 11.9 H = 13	23.	b = 9.25, B = 10.75 H = 31
8.	b = 16, B = 17 H = 20	24.	b = 27, B = 51 H = 20
9.	b = 21.4, B = 28.6 H = 12	25.	b = 28, B = 42 H = 35
10.	b = 2.2, B = 8.8 H = 8	26.	b = 4, B = 14 H = 9
11.	b = 34, B = 46 H = 11	27.	b = 11, B = 89 H = 23
12.	b = 14, B = 30 H = 17	28.	b = 14, B = 44 H = 30
13.	b = 25, B = 43 H = 30	29.	b = 16, B = 34 H = 25
14.	b = 11.3, B = 15.7 H = 40	30.	b = 15, B = 45 H = 17
15.	b = 9.9, B = 16.1 H = 12	31.	b = 15.3, B = 26.7 H = 21
16.	b = 27, B = 33 H = 14	32.	b = 3, B = 25 H = 14

Trapezoids 3

Find the base or height of the trapezoid as indicated.

1. The area of a trapezoid is 72, the height is 9, and one base is 5. Find the other base.	7. The area of a trapezoid is 165, the height is 10, and one base is 23. Find the other base.
2. The area of a trapezoid is 96, and the bases are 7 and 17. Find the height.	8. The area of a trapezoid is 90, and the bases are 9 and 36. Find the height.
3. The area of a trapezoid is 42, the height is 6, and one base is 5. Find the other base.	9. The area of a trapezoid is 290, the height is 20, and one base is 12. Find the other base.
4. The area of a trapezoid is 135, and the bases are 8 and 22. Find the height.	10. The area of a trapezoid is 625, the height is 25, and one base is 25.3. Find the other base.
5. The area of a trapezoid is 230, the height is 10, and one base is 16. Find the other base.	11. The area of a trapezoid is 286, and the bases are 10.2 and 11.8. Find the height.
6. The area of a trapezoid is 108, one base is 12, and the height is 8. Find the other base.	12. The area of a trapezoid is 167, and the bases are 13.5 and 6.5. Find the height.

Trapezoids 4

Find the perimeter in units and the area in square units of the isosceles trapezoid.

1. sides 17, 17, 17, and 33

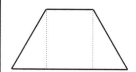

6. bases 6 and 18; legs $4\sqrt{3}$

2. bases 2 and 4; one angle 135°

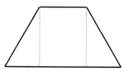

7. bases 8 and 18; one angle 120°

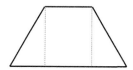

3. sides 13, 13, 13, 23

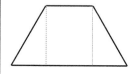

8. bases 25 and 35; one angle 45°

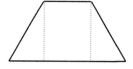

4. sides 25, 25, 25, and 39

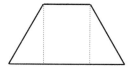

9. sides 25, 25, 30, and 60

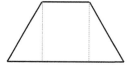

5. parallel sides 9 and 15; non-parallel sides 5

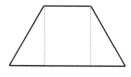

10. bases 6 and 12; one angle 60°

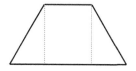

Triangles 1

Given two angles of a triangle, find the third angle by mental math.

1. 30, 60	17. 18, 100	33. 13, 99	49. 55.5, 72.5
2. 39, 51	18. 12, 12	34. 35, 53	50. 60.3, 60.6
3. 50, 50	19. 15, 115	35. 45, 85	51. 29.1, 31.9
4. 26, 90	20. 35, 45	36. 27, 57	52. 26, 134.7
5. 45, 90	21. 21, 29	37. 29, 86	53. 32.1, 43.2
6. 6, 102	22. 26, 70	38. 16, 61	54. 15.2, 91.8
7. 12, 86	23. 25, 25	39. 73, 74	55. 16.25, 83
8. 18, 36	24. 51, 52	40. 10, 55	56. 11.1, 101.1
9. 63, 100	25. 16, 101	41. 14, 84	57. 33.3, 44.4
10. 47, 90	26. 59, 60	42. 22, 75	58. 13.5, 111
11. 12, 43	27. 21, 46	43. 39, 39	59. 80.5, 81.5
12. 22, 73	28. 11, 53	44. 80, 90	60. 42.5, 49.5
13. 42, 69	29. 15, 72	45. 52, 62	61. 31.4, 52.6
14. 70, 80	30. 24, 82	46. 46, 64	62. 25.9, 75.4
15. 20, 86	31. 38, 64	47. 77, 78	63. 34.7, 43.6
16. 31, 41	32. 20, 21	48. 39, 40	64. 65.5, 69.5

Triangles 2

Identify △ABC with as many labels as possible among: Isosceles, Equilateral, Scalene, Right, Acute, Obtuse, Equiangular.

1. $AC = BC$	12. $m\angle A = 100°$, $m\angle B = 40°$
2. $AC = 11$, $BC = 11$	13. $AB = BC = AC$
3. $m\angle C = 90°$	14. $m\angle A = 33°$, $m\angle B = 33°$
4. $m\angle A = 21°$, $m\angle B = 69°$	15. $m\angle A = 51°$, $m\angle B = 51°$
5. $m\angle A = 60°$	16. $AB = 10$, $BC = 11$
6. $m\angle A = 41°$, $m\angle B = 49°$	17. $m\angle A = 90°$, $AB = 12$, $AC = 12$
7. $m\angle A = 30°$, $m\angle B = 120°$	18. $m\angle A = 40°$, $AB = 9$, $BC = 9$
8. $m\angle A = 90.01°$	19. $m\angle A = 40°$, $AB = 9$, $AC = 9$
9. $m\angle A = 60°$, $m\angle C = 60°$	20. $m\angle A = 90°$, $m\angle B = 45°$
10. $m\angle A = 41°$, $m\angle B = 52°$	21. $AB = 8$, $AC = 8$, $BC = 8\sqrt{2}$
11. $AB = 10$, $BC = 10$, $AC = 10$	22. $AB = 6$, $BC = 7$, $AC = 8$

Triangles 3

Draw 3 medians, 3 angle bisectors, and 3 altitudes in the triangles.

1.	Medians	Angle Bisectors	Altitudes

2.

3.

4.

5.

6.

7.

Triangles 4

Find the base, height, or area in square units of each triangle by mental math.

#	BASE	HEIGHT	AREA	#	BASE	HEIGHT	AREA	#	BASE	HEIGHT	AREA
1.	11	36		17.	48	220		33.	19		361
2.	29		145	18.	200	13.5		34.	80	12.5	
3.		60	1500	19.	25	180		35.	17	170	
4.	28	14		20.		11.1	555	36.	70	35	
5.		37	185	21.	220	35		37.		24	144
6.	50		1250	22.	40	143		38.		16	256
7.	12	97		23.	14		175	39.	13.5	80	
8.	20	50		24.	22	59		40.	25		675
9.	15		225	25.	45	440		41.	45	90	
10.		16	96	26.		15	450	42.	26	26	
11.	14	42		27.	120	140		43.	70	75	
12.	11	78		28.	240		1320	44.		22	594
13.	40		800	29.		200	8600	45.	20.2	30	
14.		15	60	30.	80	55		46.	24.5	40	
15.	18	18		31.		121	2420	47.		25	900
16.		36	396	32.	22		660	48.		65	1950

Triangles 5

Find the area. One box equals one square unit.

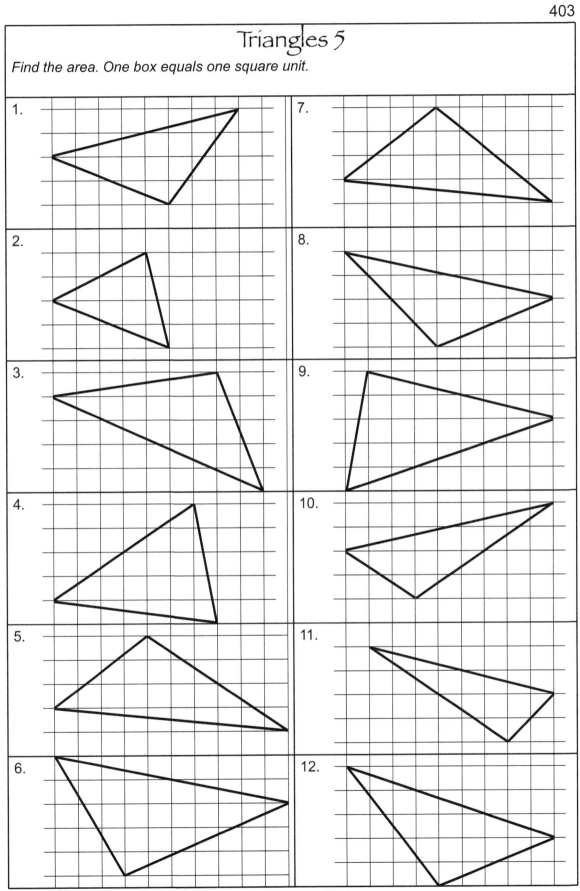

Triangles 6

Given the measures for three sides, answer yes or no as to whether the numbers form a triangle.

1. 8, 8, 8	17. 9, 12, 25	33. 13, 13, 28
2. 3, 5, 9	18. 7, 18, 21	34. 18, 18, 35.9
3. 3, 4, 5	19. 3.2, 5.8, 8.1	35. 2, 4, 8
4. 2, 3, 5	20. .01, .01, .01	36. 16, 18, 24
5. 12, 25, 40	21. 15, 20, 25	37. 11, 11, 11
6. 11, 13, 24	22. 16, 23, 39	38. 6, 7, 12
7. 8.2, 8.2, 8.2	23. 11, 12, 13	39. 15, 15, 25
8. 7, 15, 21	24. 19, 22, 35	40. 9, 10, 18.7
9. 6, 7, 8	25. 13.1, 22, 35	41. 22, 23, 44
10. 400, 600, 1000	26. 10, 14, 23.9	42. 31, 33, 54
11. 15, 20, 32	27. 33, 34, 43	43. 20, 24, 40
12. 2.5, 3.5, 4.5	28. 13.1, 13.2, 26.3	44. 17, 21, 36
13. 13, 13.1, 13.2	29. 19, 33, 51	45. 1.5, 5.5, 7.5
14. 14, 16, 30	30. 50, 50, 51	46. 1, 1, 3
15. 2, 10, 11	31. 46, 34, 90	47. 18, 22, 37
16. 4.4, 5.5, 10	32. 48, 33, 80	48. 37, 39, 75

Triangles 7

Given two sides of a triangle,	write an inequality for the value of the 3rd side s.	write an inequality for the value of the perimeter P.
1. 3, 12		
2. 9, 13		
3. 6, 15		
4. 8, 20		
5. 12, 13		
6. 8, 18		
7. 3, 4		
8. 12, 15		
9. 11, 22		
10. 5, 17		
11. 15, 32		
12. 14, 19		
13. 31, 38		
14. 22, 45		
15. 20, 43		
16. 18, 21		
17. 10, 31		
18. 16, 30		
19. a, b; a < b		

Triangles 8

Answer as indicated, making a list when necessary.	Answer as indicated. Linear measures are units; area measures are square units.
1. How many triangles with whole number sides have longest side 5?	7. Find the area of an isosceles right triangle with hypotenuse 10.
2. How many triangles with whole number sides have longest side 6?	8. Find the leg of an isosceles right triangle with area 32.
3. Given two sticks with lengths 3 and 11. Find all possible integral lengths of a third stick that could be used to construct a triangle.	9. The area of a rectangle 9 by 6 equals the area of a triangle with base 27. Find the corresponding altitude of the triangle.
4. A triangle has one side 3 and another side 7. Find the product of the least and greatest possible integral values of the perimeter.	10. Find the leg of an isosceles right triangle with area 50.
5. A triangle has one side 4 and another side 13. Find the sum of the least and greatest possible integral values of the perimeter.	11. If the area of a square with side 18 equals the area of a triangle with altitude 8, find the corresponding base of the triangle.
6. Given two sticks with lengths 5 and 12. How many possible integral lengths of a third stick could be used to construct a triangle?	12. Find the area of an isosceles triangle with perimeter 36 and altitude to the base 12.

Triangles 9

Answer as indicated. NTS

1. Find x.

6. Find x + y.

2. Find x + y.

7. Find x.

3. Find x.

8. If the largest triangle is equilateral, find y.

4. Find x.

9. Find x if the triangle is isosceles with a 75° vertex angle.

5. Find x given the smaller bottom triangle is isosceles with a 37° base angle.

10. Find x.

Triangles 10

Answer YES or NO as to whether the triangles are similar.

1. all right triangles	17. all 20°-60°-100° triangles
2. all scalene triangles	18. all isosceles triangles with base 5
3. all obtuse triangles	19. all isosceles triangles with base angles 20°
4. all equilateral triangles	20. all right triangles with hypotenuse 10
5. all isosceles right triangles	21. all right triangles with a leg of 10
6. all equiangular triangles	22. all isosceles right triangles with a leg of 10
7. all isosceles triangles	23. all rotations of 6-6-7 triangles
8. all acute triangles	24. all isosceles triangles with congruent sides 15
9. all 30°-60°-90° triangles	25. all isosceles triangles with two angles 60°
10. all 45°-45°-90° triangles	26. all triangles with a 70° angle and a side 10
11. all triangles with a 100° angle	27. all 25°-40°-115° triangles
12. all obtuse triangles with a 120° angle	28. all dilations of 5-5-5 triangles
13. all right triangles with a 10° angle	29. all scalene triangles with 2 sides 10 and 12
14. all dilations of 3-4-5 triangles	30. all acute triangles with a 40° angle
15. all isosceles triangles with a vertex angle 20°	31. all dilations of 5-6-7 triangles
16. all isosceles triangles with two sides 10	32. all reflections of 4-6-8 triangles

Triangles 11

Complete the chart for 2 similar triangles.

GIVEN	RATIO OF SIDES	RATIO OF ALTITUDES	RATIO OF PERIMETERS	RATIO OF AREAS
1. Perimeters 5 and 9				
2. Altitudes 3 and 7				
3. Sides 4, 8, 10 and 6, 12, 15				
4. Greatest sides 12 and 15				
5. Sides 3, 4, 5 and 18, 24, 30				
6. Greatest sides 15 and 20				
7. Altitudes 4 and 12				
8. Perimeters 12 and 20				
9. Least sides 16 and 40				
10. Sides 5, 12, 13 and 10, 24, 26				
11. Perimeters 14 and 22				
12. Least sides 18 and 21				
13. Greatest sides 30 and 42				
14. Altitudes 11 and 44				
15. Perimeters 10 and 45				
16. Least sides 12 and 44				
17. Altitudes 24 and 27				
18. Greatest sides 21 and 36				
19. Perimeters 12 and 45				

Triangles 12

△ABC ~ △DEF. Find the lengths of the two sides not given using mental math. NTS

1. AB = 5, BC = 3, AC = 4, ED = 10

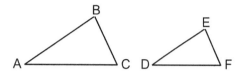

7. AB = 6, AC = 9, BC = 8, EF = 12

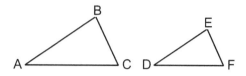

2. ED = 8, EF = 6, DF = 4, AC = 6

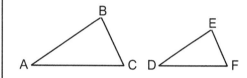

8. AB = 3, BC = 4, AC = 5, DF = 20

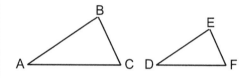

3. AB = 6, BC = 9, AC = 12, EF = 3

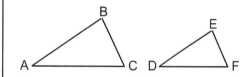

9. AC = 10, BC = 6, DE = 6, DF = 12

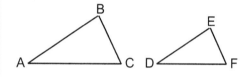

4. ED = 6, EF = 9, DF = 9, AB = 8

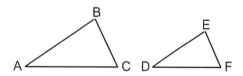

10. AB = 9, AC = 12, DE = 15, EF = 30

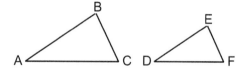

5. AB = 9, BC = 18, AC = 12, DF = 8

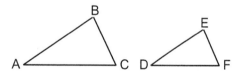

11. AB = 14, DF = 12, DE = 6, EF = 9

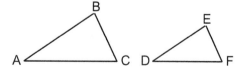

6. AB = 8, AC = 12, BC = 10, DE = 6

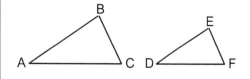

12. DE = 6, EF = 18, AB = 8, AC = 20

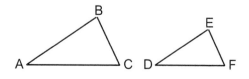

Triangles 13

Find x. NTS	Find the area of the triangle by Hero.
1.	8. 4, 6, 8
2.	9. 3, 6, 7
3.	10. 5, 7, 10
4.	11. 4, 4, 6
5.	12. 3, 7, 8
6.	13. 13, 14, 15
7.	14. 5, 6, 7

Triangles 14

Find the missing sides, area, and perimeter. NTS

$m \angle A = 30°.$	$m \angle A = 45°.$
1. 12	8. 12
2. 12	9. 12
3. 12	10. 10
4. $\sqrt{3}$	11. 60, 60°
5. $8\sqrt{3}$	12. $\sqrt{30}$
6. 18	13. 60, 60°
7. 8	14. $\sqrt{2}$

MAVA Math: Enhanced Skills Copyright © 2015 Marla Weiss

Triangles 15

Find the area in square units of the isosceles triangle with the given sides.

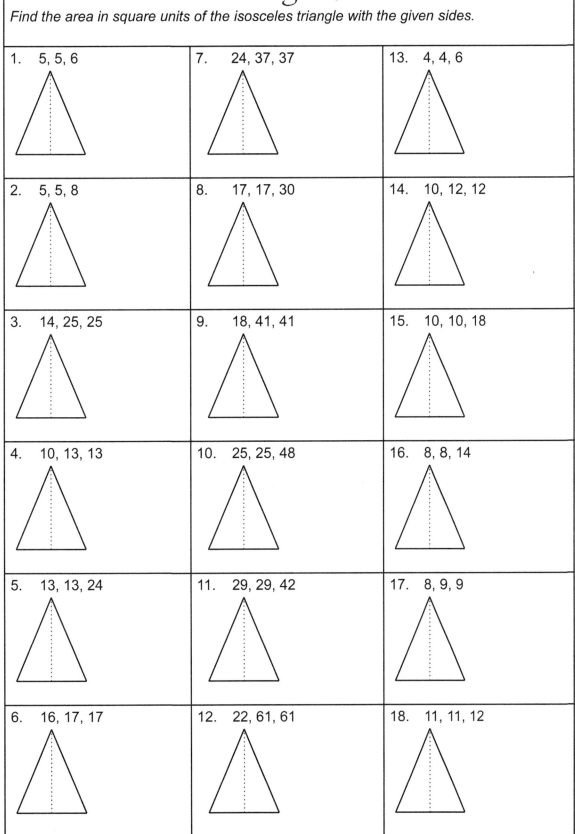

1. 5, 5, 6

2. 5, 5, 8

3. 14, 25, 25

4. 10, 13, 13

5. 13, 13, 24

6. 16, 17, 17

7. 24, 37, 37

8. 17, 17, 30

9. 18, 41, 41

10. 25, 25, 48

11. 29, 29, 42

12. 22, 61, 61

13. 4, 4, 6

14. 10, 12, 12

15. 10, 10, 18

16. 8, 8, 14

17. 8, 9, 9

18. 11, 11, 12

Triangles 16

Find the altitude to the hypotenuse.	Find the median to the hypotenuse.
1. sides 3, 4, 5	9. sides 3, 4, 5
2. sides 5, 12, 13	10. sides 5, 12, 13
3. sides 7, 24, 25	11. sides 7, 24, 25
4. sides 8, 15, 17	12. sides 8, 15, 17
5. sides 9, 40, 41	13. sides 9, 40, 41
6. sides 11, 60, 61	14. sides 11, 60, 61
7. sides 12, 35, 37	15. sides 12, 35, 37
8. sides 20, 21, 29	16. sides 20, 21, 29

Triangles 17

Find the area of the equilateral triangle with the given side.	Find the area of the regular hexagon with the given side.
1. s = 2	14. s = 3
2. s = 4	15. s = 6
3. s = 5	16. s = 7
4. s = 8	17. s = 9
5. s = 10	18. s = 11
6. s = 12	19. s = 18
7. s = 14	20. s = 22
8. s = 16	21. s = 26
9. s = 20	22. s = 28
10. s = 24	23. s = 30
11. s = 32	24. s = 36
12. s = 40	25. s = 42
13. s = 44	26. s = 48

Triangles 18

Answer as indicated for the congruent triangles. NTS

Answer as indicated for adjacent or overlapping triangles. NTS

1. $\triangle ACB \cong \triangle ACD$. Find the perimeter of quadrilateral ABCD.

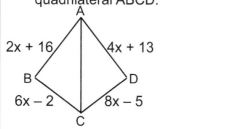

6. Find the area and perimeter of quadrilateral ABCD.

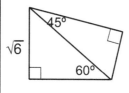

2. $\triangle ACB \cong \triangle DCE$. AC = 2x + 9. BC = 15. CD = 4x − 1. Find AE.

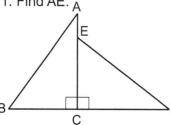

7. Given AE = 9, AC = 12, F is one of the two trisection points of $\overline{AE}$ closer to E, and B is the midpoint of $\overline{AC}$. Find the sum of the areas of $\triangle ACF$ and $\triangle ABE$.

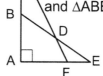

3. $\triangle ACB \cong \triangle DBC$. Find the greatest possible whole sum of the perimeters of the 2 triangles.

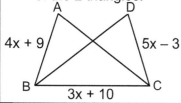

8. Find the perimeter of the pentagon.

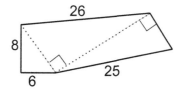

4. $\triangle ABC \cong \triangle DEF$. Given A(−6,0), B(0,−5), C(0,0), D(2,4), and F(2,10). Find the sum of all x and y values of E(x,y).

9. If all small squares are congruent, the area of the total shaded regions is what fractional part of the area of the greatest square?

5. $\triangle ABC \cong \triangle DEF$. AB = 3x − 7. BC = 8. AC = 5z + 2. DE = 11. DF = 12. EF = 3y + 11. Find x − y + z.

10. If the area of $\triangle ABC$ is 48, what is the area of $\triangle ACD$?

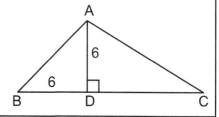

Triangles 19

Answer as indicated for the similar triangles. NTS

---

1. The area of the larger triangle is 96. Find the horizontal base of the smaller triangle.

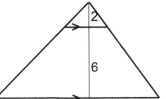

---

6. Find the area of the right trapezoid.

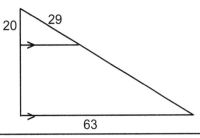

---

2. Find the sum of the perimeters of the 2 triangles.

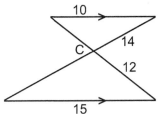

---

7. A, B, C, D are midpoints of sides of a parallelogram with diagonals 20 and 30. Find the perimeter of ABCD.

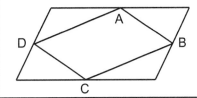

---

3. Find x.

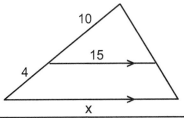

---

8. △ABC ~ △EBD.
BC = 30. BD = 20.
BE = 30. Find AE.

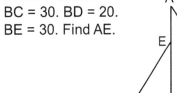

---

4. Find the perimeter of the larger triangle.

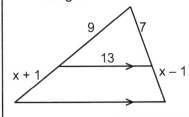

---

9. Find the perimeter of the smallest triangle.

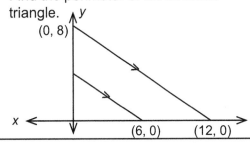

---

5. The horizontal lines are parallel. Find x, y, and z if their sum is 25.

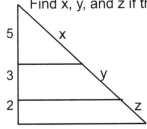

---

10. Find the measures of the 2 missing sides.

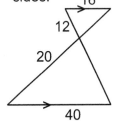

MAVA Math: Enhanced Skills Copyright © 2015 Marla Weiss

Triangles 20

Find the probability that 3 numbers selected randomly from a list of numbers form the sides of a triangle.

1. 3, 5, 8, and 10	9. 2, 6, 8, 9, and 11
2. 3, 5, 7, and 9	10. 2, 3, 5, 8, and 9
3. 2, 3, 5, and 7	11. 4, 6, 7, 8, and 9
4. 4, 6, 8, and 10	12. 3, 6, 7, 9, and 15
5. 4, 5, 7, and 9	13. 3, 4, 5, 8, and 11
6. 3, 5, 7, and 10	14. 1, 3, 5, 9, and 13
7. 2, 3, 5, and 6	15. 2, 4, 5, 7, and 11
8. 3, 5, 6, and 8	16. 4, 5, 8, 9, and 13

Triangles 21

Given triangles with sides: 3, 4, 5; 3, 6, 8; 6, 6, 6; 6, 7, 7; 6, 7, 8; 6, 8, 10; and 8, 8, 8.
Selecting two of the triangles, find the probability that:

WORK SPACE

1. the triangles are similar.	8. the triangles are both isosceles.
2. the triangles are both right.	9. the triangles are congruent.
3. the triangles are both acute.	10. all 6 altitudes are in the interiors of the triangles.
4. the triangles are both obtuse.	11. the triangles are both scalene.
5. the triangles are both equilateral.	12. exactly 4 of the 6 altitudes are the sides.
6. one triangle is right and one is acute.	13. one triangle is equilateral and one is scalene.
7. one triangle is right and one is obtuse.	14. one triangle is acute and one is obtuse.

Triangles 22

Find the probability that 3 numbers selected randomly form the sides of a right △.	Find the probability that 3 numbers selected randomly form the angles of a right triangle.
1. 3 to 13 inclusive	9. 10, 30, 40, 50, 60, 80, and 90
2. 7 to 26 inclusive	10. 15, 20, 40, 50, 75, and 90
3. 5 to 20 inclusive	11. 25, 35, 55, 65, and 90
4. 6 to 17 inclusive	12. 5, 15, 20, 60, 70, 75, 85, and 90
5. 15 to 29 inclusive	13. 5, 10, 15, 20, 70, 75, 80, 85, and 90
6. 12 to 37 inclusive	14. 12, 33, 39, 57, 61, 78, and 90
7. 20 to 29 inclusive	15. 3, 16, 17, 18, 63, 72, 74, 87, and 90
8. 5 to 25 inclusive	16. 4, 13, 21, 46, 59, 77, 86, and 90

Triplets 1

Answer using the "keep 1–change 2" chart method. Assume uniform rates.

1. If 6 cats can catch 6 mice in 6 minutes, how many cats are needed to catch 100 mice in 100 minutes?	5. Ten movers can pack 15 studios in 2 days. How many movers are needed to pack 42 studios in one week?
2. Five boys can pack 60 boxes in 3 days. How many boxes can 2 boys pack in 4 days? boys	6. Nine chefs can prepare 51 dinners in 2 hours. 24 chefs can prepare how many dinners in 1 day?
3. Four girls can mow 30 lawns in 5 days. Eight girls can mow how many lawns in a day-and-a half?	7. If 8 cats can catch 12 mice in 20 minutes, how many cats are needed to catch 36 mice in 2 hours?
4. 2000 bees need one year to make 7 pounds of honey. How many years do 5000 bees need to make 70 pounds of honey?	8. If a chicken-and-a-half lays an egg-and-a-half in a day-and-a-half, how many eggs will 12 chickens lay in 12 days?

Triplets 2

Answer using the "keep 1–change 2" method. Assume uniform rates and positive, integral variables.

1.

If *x* students can drink *y* cans of soda in *z* days, how many cans of soda can *w* students drink in 1 week?

2.

If *c* cats can catch *m* mice in a half-hour, how many cats are needed to catch *e* mice in 1 week?

3.

If *c* chickens lay *e* eggs in *d* days, how many eggs will *k* chickens lay in June?

4.

If *b* boys can pack *x* boxes in *h* hours, how many boxes can *y* boys pack in one day?

5.

If *m* men can mow *w* lawns in 1 week, *n* men can mow how many lawns in 52 weeks?

6.

If *e* chefs can prepare *m* meals in 1 week, *f* chefs can prepare how many meals in 1 hour?

7.

If *v* movers can pack *r* rooms in *d* days, how many movers are needed to pack *m* rooms in January?

8.

If *b* bees need one year to make *p* pounds of honey, *e* bees will make how many pounds of honey in one century?

Vocabulary 1

Complete the blank.

1. In a plane, two lines that are _____ form four right angles.

2. The x-axis and y-axis divide the coordinate plane into four _____.

3. A line that crosses two parallel lines is called a _____.

4. If a=b and b=c, then a=c by the property _____ of equality.

5. An angle with its vertex at the center of a circle is called a _____ angle.

6. A ratio equal to another ratio is called a _____.

7. A point that bisects a line segment is called its _____.

8. A _____ fraction has fractions in its numerator and/or denominator.

9. The _____ of sets is the set of elements common to the given sets.

10. A synonym for element of a set is _____.

11. The more common name for multiplicative inverse is _____.

12. An angle measuring 180º is called a _____.

13. The longest chord of a circle is a _____.

14. The union of two _____ sharing a common endpoint is an angle.

15. A regular quadrilateral is called a _____.

16. The more common name for arithmetic mean is _____.

17. The tilt of a line is called its _____.

18. On the coordinate plane, a _____ has integers as both coordinates.

19. A synonym for height of a triangle is _____.

Vocabulary 2

Complete the blanks.

1. An isosceles triangle has one _____ angle and two _____ angles.

2. "Positive, _____, or zero" is an example of a _____.

3. 2 angles are _____ if their degree sum is 90° but _____ if 180°.

4. The distributive field property uses the operations _____ and _____.

5. 5 and 7 are _____ primes; 7 and 11 are _____ primes.

6. A _____ number has 3 or more factors; a _____ number has exactly 2 different factors.

7. Formed by parallel lines and a transversal, alternate angles may be _____ or _____.

8. A triangle with 3 congruent sides is both _____ and _____.

9. Dividing by _____ and raising _____ to the 0th power are undefined.

10. 1, 2, 3, 4, . . . are called _____ or _____ numbers.

11. Circumference is to _____ as _____ is to polygon.

12. The identity element for multiplication is _____ and for addition is _____.

13. _____, _____, and mode are three measures of central tendency.

14. A set of ordered pairs is a _____, which may or may not be a _____.

15. In a parallelogram, _____ angles are supplementary, and _____ angles are congruent.

16. The sides of a right triangle are called the _____ and the _____.

17. Three set operations are _____, _____, and complementation.

18. 1, 4, 9, 16, . . . are _____; 1, 8, 27, 64, . . . are _____.

19. The 2 parts of a fraction are its _____ and _____.

Work Problems 1

Find the time needed for the people to complete the job working together.

1. Person A can do the job in 1 hour. Person B can do the job in 2 hours.	8. Person A can do the job in 5 hours. Person B can do the job in 7 hours.
2. Person A can do the job in 1 hour. Person B can do the job in 3 hours.	9. Person A can do the job in 7 hours. Person B can do the job in 8 hours.
3. Person A can do the job in 2 hours. Person B can do the job in 3 hours.	10. Person A can do the job in 6 hours. Person B can do the job in 9 hours.
4. Person A can do the job in 2 hours. Person B can do the job in 4 hours.	11. Person A can do the job in 4 hours. Person B can do the job in 6 hours.
5. Person A can do the job in 1 hour. Person B can do the job in 4 hours.	12. Person A can do the job in 1 hour. Person B can do the job in 5 hours.
6. Person A can do the job in 2 hours. Person B can do the job in 6 hours.	13. Person A can do the job in 5 hours. Person B can do the job in 5 hours.
7. Person A can do the job in 3 hours. Person B can do the job in 6 hours.	14. Person A can do the job in 4 hours. Person B can do the job in 8 hours.

Work Problems 2

Find the time for Person B alone to do the job given the time for A alone to do the job and their time together.

1.   Time for Person A alone = 4 hours
     Time together to finish = 3 hours
	7. Time for Person A alone = 4 hours
	     Time together to finish = 1.5 hours

2.   Time for Person A alone = 4 hours
     Time together to finish = 2 hours
	8. Time for Person A alone = 5.5 hours
	     Time together to finish = 3 hours

3.   Time for Person A alone = 3 hours
     Time together to finish = 1 hour
	9. Time for Person A alone = 4 hours
	     Time together to finish = 1 hour

4.   Time for Person A alone = 4 hours
     Time together to finish = 2.5 hours
	10. Time for Person A alone = 5 hours
	      Time together to finish = 3.5 hours

5.   Time for Person A alone = 3.5 hours
     Time together to finish = 2.5 hours
	11. Time for Person A alone = 6 hours
	      Time together to finish = 4.5 hours

6.   Time for Person A alone = 5 hours
     Time together to finish = 3 hours
	12. Time for Person A alone = 7 hours
	      Time together to finish = 3 hours

ABBREVIATIONS

Use common sense in decoding abbreviations. For example, in division problems, R means remainder, yet R means radius in geometry and common ratio in sequences.

General

CBD	can't be done
DS	digit sum
GCF	greatest common factor
LCM	least common multiple
NTS	not to scale
T&E	trial and error
WP	word problem
WLOG	without loss of generality

Coordinate Plane

D	down
L	left
Q	quadrant
R	right
U	up

Logic

T	true
F	false

Sets

N	Natural numbers
W	Whole numbers
Z	Integers
Q	Rational numbers
R	Real numbers
C	Complex numbers

Geometry

A	area
B	base
C	circumference
D	diameter
E	edge
H	height
L	length
M	midline of a trapezoid
P	perimeter
R	radius
S	side
SA	surface area
V	volume
W	width

Functions

ABS	absolute value
SGN	1 for > 0, –1 for < 0, 0 for = 0
TRUNC	truncate decimal part
INT	greatest integer contained in
SIGMA	number of divisors
SQR	square
SQRT	square root

Money

D	dime
H	half dollar
N	nickel
P	penny
Q	quarter

Percent Change

D	decrease
I	increase

Triangles

A	acute
E	equilateral
I	isosceles
O	obtuse
R	right
S	scalene

Solids

E	number of edges
F	number of faces
V	number of vertices

Sequences

A	arithmetic
d	common difference
G	geometric
r	common ratio

Word Problems

D	distance
R	rate
T	time

Arithmetic

R	remainder

ABBREVIATIONS (continued)

<u>Properties</u>

ClPA	Closure Property of Addition
ClPM	Closure Property of Multiplication
APA	Associative Property of Addition
APM	Associative Property of Multiplication
CPA	Commutative Property of Addition
CPM	Commutative Property of Multiplication
IdPA	Identity Property of Addition
IdPM	Identity Property of Multiplication
InPA	Inverse Property of Addition
InPM	Inverse Property of Multiplication
DPMA	Distributive Property of Multiplication over Addition
ZPM	Zero Property of Multiplication
RPE	Reflexive Property of Equality
SPE	Symmetric Property of Equality
TPE	Transitive Property of Equality
APE	Addition Property of Equality
MPE	Multiplication Property of Equality
APU	Associative Property of Union
CPU	Commutative Property of Union
API	Associative Property of Intersection
CPI	Commutative Property of Intersection
DPUI	Distributive Property of Union over Intersection
DPIU	Distributive Property of Intersection over Union
RPC	Reflexive Property of Congruence
SPC	Symmetric Property of Congruence
TPC	Transitive Property of Congruence
APC	Addition Property of Congruence
MPC	Multiplication Property of Congruence

NOTES

FORMULAE

Area square s^2	Diagonal square $s\sqrt{2}$
Area rectangle bh	Diagonal cube $e\sqrt{3}$
Area rhombus $Dd/2$	Diagonal rectangular prism $\sqrt{a^2+b^2+c^2}$
Area parallelogram bh	sum of the angles of an n-gon $180(n-2)$
Area trapezoid $h(b+B)/2$	number of diagonals of an n-gon $n(n-3)/2$
Area triangle $bh/2$	interior angle of a regular n-gon $180(n-2)/n$
Area equilateral triangle $\dfrac{s^2\sqrt{3}}{4}$	exterior angle of a regular n-gon $360/n$
Area circle πr^2	midpoint of (x, y) and (w, z) $((x+w)/2,(y+z)/2)$
Perimeter square $4s$	Pythagorean Theorem $a^2+b^2=c^2$
Perimeter rectangle $2(L+W)$	GCF(m,n) x LCM(m,n) = mn
Circumference circle πd	distance = rate x time
Surface area cube $6e^2$	simple interest = principal x rate x time
Surface area rectangular prism $2(ab+bc+ac)$	value after compound interest $P(1+r)^T$
Surface area sphere $4\pi r^2$	number of subsets of a set with n elements 2^n
Surface area cylinder $2\pi r^2+\pi dh$	number of squares in an n by n grid $1^2+2^2+3^2+\ldots+n^2$
Surface area cone $\pi r^2+\pi r\ell$	number of rectangles in an n by n grid $(1+2+3+\ldots+n)^2$
Volume cube e^3	
Volume rectangular prism lwh	E + F + V = 6N + 2 | for prisms:

Volume right prism Bh	E = 3N | N = # sides of base
	E = # edges

Volume pyramid $Bh/3$	F = N + 2 | V = # vertices
	F = # faces

Volume cylinder πr^2h	V = 2N |
Volume cone $\pi r^2h/3$	Hero: Area of Triangle $\sqrt{S(S-a)(S-b)(S-c)}$ S = semi-perimeter
Volume sphere $4\pi r^3/3$	Pick: Area on Geoboard $B/2+I-1$ B = # border points, I = # interior pts

48 PYTHAGOREAN TRIPLES

3, 4, 5
5, 12, 13
7, 24, 25
8, 15, 17
9, 40, 41
11, 60, 61
12, 35, 37
13, 84, 85
15, 112, 113
16, 63, 65
17, 144, 145
19, 180, 181
20, 21, 29
20, 99, 101
21, 220, 221
23, 264, 265
24, 143, 145
28, 45, 53
28, 195, 197
32, 255, 257
33, 56, 65
35, 612, 613
36, 77, 85
39, 80, 89
44, 117, 125
48, 55, 73
51, 140, 149
52, 165, 173
57, 176, 185
60, 91, 109
60, 221, 229
65, 72, 97
68, 285, 293
69, 260, 269
84, 187, 205
85, 132, 157
88, 105, 137
95, 168, 193
96, 247, 265
104, 153, 185
105, 208, 233
115, 252, 277
119, 120, 169
120, 209, 241
133, 156, 205
140, 171, 221
160, 231, 281
161, 240, 289

PRIME NUMBERS
(Less Than 1070)

2	199	467	769
3	211	479	773
5	223	487	787
7	227	491	797
11	229	499	809
13	233	503	811
17	239	509	821
19	241	521	823
23	251	523	827
29	257	541	829
31	263	547	839
37	269	557	853
41	271	563	857
43	277	569	859
47	281	571	863
53	283	577	877
59	293	587	881
61	307	593	883
67	311	599	887
71	313	601	907
73	317	607	911
79	331	613	919
83	337	617	929
89	347	619	937
97	349	631	941
101	353	641	947
103	359	643	953
107	367	647	967
109	373	653	971
113	379	659	977
127	383	661	983
131	389	673	991
137	397	677	997
139	401	683	1009
149	409	691	1013
151	419	701	1019
157	421	709	1021
163	431	719	1031
167	433	727	1033
173	439	733	1039
179	443	739	1049
181	449	743	1051
191	457	751	1061
193	461	757	1063
197	463	761	1069

Middle School Math Vocabulary

absolute value
abundant number
acute angle
acute triangle
add
addition
Addition Property of Equality
additive inverse
adjacent
adjacent angles
adjacent sides
age problem
algebra
algebraic expression
algorithm
alternate exterior angles
alternate interior angles
altitude
AND
angle
apothem
approximation
arc
area
arithmetic
arithmetic mean
arithmetic sequence
array
ascending order
Associative Property of Addition
Associative Property of Multiplication
at least
at most
average
average rate
axis (axes)
bar graph
base
billion
binary
bisect
bisection
boundary problem

calculator
calendar problem
center
centimeter
central angle
chart
chord
cipher
circle
circle graph
circumference
circumscribed
Closure Property of Addition
Closure Property of Multiplication
coefficient
coin problem
collinear
combination
common denominator
common difference
common element
common factor
common fraction
common multiple
common ratio
Commutative Property of Addition
Commutative Property of Multiplication
compass
complement
complementary angles
complementation
complex fraction
complex number
composite number
composition
compound interest
concatenation
concave
concentric
cone
congruent
consecutive
consecutive even

consecutive odd	divisor
constant	dodecagon
convex	dodecahedron
coordinate plane	domain
coordinates	double
coplanar	edge
corresponding angles	element
corresponding sides	empty set
counterexample	endpoint
counting number	equal
cross multiply	equality
cube	equation
cube root	equiangular
cylinder	equidistant
data	equilateral
decagon	equilateral triangle
decimal	equivalent
decimeter	equivalent fractions
decrease	estimation
decrement	Euler Graph
definition	evaluate
degree	even number
denominator	event
dependent events	expanded notation
descending order	exponent
diagonal	exponential form
diameter	exponentiation
dichotomy	expression
difference	exterior
digit	exterior angle
dimension	externally tangent
direct variation	face
directed graph	factor
discount	factorial
discrete sets	fence post problem
disjoint sets	Fibonacci Sequence
distance	fictitious operation
distance-rate-time problem	field
distinct	First-Plus-Last Method
Distributive Property	flip
dividend	foot (feet)
divisibility rule	formula
divisible	fraction
division	fractional part

frequency distribution	inscribed
frequency table	insufficient information
function	integer
geometric mean	intercept
geometric sequence	interest
geometry	interior
Goldbach's Conjecture	interior angle
graph	internally tangent
greater	interpolate
greater than	intersect
greater than or equal to	intersecting lines
greatest	intersection
greatest common divisor	inverse
greatest common factor	inverse operation
half	Inverse Property of Addition
halve	Inverse Property of Multiplication
height	inverse variation
hemisphere	irrational number
heptagon	isosceles right triangle
hexagon	isosceles trapezoid
hexagonal prism	isosceles triangle
hexagonal pyramid	kilometer
histogram	kite
How Many? problem	lateral area
hundred	lateral edge
hundredth	lattice point
hypotenuse	least
i	least common denominator
icosahedron	least common multiple
identity element	leg
Identity Property of Addition	length
Identity Property of Multiplication	less
imaginary number	less than
improper fraction	less than or equal to
in terms of	like terms
inch	line
increase	line of symmetry
increment	line segment
independent events	linear equation
inequality	linear function
infinite	long division
infinite sequence	lowest common denominator
infinite series	lowest terms
infinity	magic square

mathematical maturity	number line
mathematician	number theory
mathematics	numeral
matrix (matrices)	numerator
matrix addition	obtuse angle
matrix multiplication	obtuse triangle
maximum	octagon
mean	octahedron
measure	odd number
measurement	odds
median	ones digit
member	ones place
mental math	operation
meter	opposite
metric system	opposite angles
midline	opposite sides
midpoint	OR
millimeter	order
million	ordered pair
minimum	organized list
mixed number	origin
mixture problem	original cost
mode	outcome
modulo	outlier
multiple	palindrome
multiplication	parabola
Multiplication Principle	parallel
Multiplication Property of Equality	parallelepiped
multiplicative inverse	parallelogram
mutually exclusive	parenthesis (parentheses)
natural number	pattern
negate	pentagon
negative number	pentagonal prism
negative sign	pentagonal pyramid
net	percent
nonagon	percent change
noncollinear	perfect cube
nonoverlapping	perfect number
nonzero	perfect square
NOT	perimeter
not to scale	permutation
nth term	perpendicular
number	perpendicular bisector
number base	perpendicular lines

perpendicular planes	ratio
pi	rational
place	ray
place value	real number
plane	reciprocal
point	rectangle
polygon	rectangular prism
polyhedron	rectangular pyramid
positive number	rectangular solid
power	reflection
power of ten	Reflexive Property of Equality
power of two	region
preceding term	region bounded by
prime factorization	regular
prime number	regular polygon
principal	relation
prism	relatively prime
probability	remainder
problem solving	repeating decimal
product	rhombus
profit problem	right angle
proof	right cylinder
proper fraction	right prism
proper subset	right pyramid
property	right triangle
proportion	rigid motion
protractor	root
prove	rotation
pyramid	sale price
Pythagorean primitive or triplet	sales tax
Pythagorean Theorem	scalar multiplication
Quadrant I, II, III, and IV	scalene trapezoid
quadratic equation	scalene triangle
quadratic formula	scientific notation
quadrilateral	sector
quadruple	segment
quantity	semicircle
quarter	semiperimeter
quotient	sequence
radical	series
radius	set
random number	shaded area
range	short division
rate	side

similar	term
similar polygons	terminating decimal
similar triangles	tetrahedron
simple interest	thousand
simplest form	thousandth
simplify	transformation
skew lines	Transitive Property of Equality
slant height	transitivity
slide	translation
slope	transversal
solid	trapezoid
solution	triangle
solution set	trichotomy
solve	trillion
space	trillions place
special right triangle	triple
sphere	triplet problem
square	trisect
square root	trisection
standard form	twin primes
standard notation	undefined
statistics	uniform border problem
stem-and-leaf plot	union
straight angle	unit
straight edge	unit fraction
subset	units digit
substitution	units place
subtraction	universal set
succeeding term	value
successive terms	variable
sum	Venn Diagram
sum of the digits	vertex (vertices)
supplement	vertical angles
supplementary angles	volume
surface area	whole number
symbol	width
symmetry	work problem
symmetric about	x-axis
Symmetric Property of Equality	x-coordinate
system of equations	yard
tangent	y-axis
ten billions place	y-coordinate
ten millions place	zero
tens place	Zero Property of Multiplication

Printed in the United States
By Bookmasters